NUREG-1863

Review of Responses to NRC Bulletin 2003-02-Leakage from Reactor Pressure Vessel Lower Head Penetrations and Reactor Coolant Pressure Boundary Integrity

Manuscript Completed: August 2006
Date Published: September 2006

Prepared by
E. Sullivan and G. Cheruvenki

Division of Component Integrity
Office of Nuclear Reactor Regulation
U.S. Nuclear Regulatory Commission
Washington, DC 20555-0001

ABSTRACT

Cracking in pressurized water reactor (PWR) bottom mounted instrumentation (BMI) fabricated from Alloy 600 base material was first identified at the South Texas Project, (STP) Unit 1 plant in the United States (US). Based on the failure analysis of the BMI, the licensee concluded that the cracking was due to primary water stress corrosion cracking (PWSCC). PWSCC has been identified as the primary degradation mechanism affecting PWR high nickel alloy nozzles and welds (e.g., Alloy 600 tubing, piping, or forging material, and Alloy 82/182 weld material) in the reactor coolant system. To address these concerns, the Nuclear Regulatory Commission (NRC) issued NRC Bulletin 2003-02, "Leakage from Reactor Pressure Vessel Lower Head Penetrations and Reactor Coolant Pressure Boundary Integrity," on August 21, 2003, to all holders of operating licenses for PWRs. The purpose of the bulletin was to request information from the industry related to the structural integrity of the reactor pressure vessel BMI nozzles at PWR facilities. This report summarizes the NRC staff's review of licensee responses to the Bulletin, licensee's BMI inspection results, industry activities related to BMI inspections, and the staff's conclusions regarding the need for additional regulatory action in this area. A brief summary regarding the inspection results of the BMI penetrations and the associated aging monitoring programs for the foreign reactors is included in this report.

CONTENTS

Figures

List of Tables

EXECUTIVE SUMMARY

The reactor pressure vessel (RPV) lower head and bottom mounted instrumentation (BMI) penetrations at South Texas Project, Unit 1 (STP-1) were visually inspected on April 12, 2003, as a routine part of the unit's refueling outage. The bare-metal visual (BMV) inspection found small amounts of white residue around two of the 58 BMI penetrations at the junction where the penetrations met the lower reactor vessel head. Based on a chemical analysis of the residue, the licensee concluded that it was boric acid from the reactor coolant. The licensee performed a destructive examination of one of the penetrations and determined that there was a through-wall flaw due to primary water stress corrosion cracking (PWSCC). PWSCC has been identified as the primary degradation mechanism affecting PWR high nickel alloy nozzles and welds (e.g., Alloy 600 tubing, piping, or forging material, and Alloy 82/182 weld material) in the reactor coolant system. The staff concluded that cracking at STP-1 and the small amount of leakage from the cracks did not represent an immediate safety problem due to the size and axial orientation of the cracks. However, the STP-1 experience demonstrated that BMV inspection of BMI penetrations is a useful inspection technique for detecting minor leakage and may assist in detecting flaws before they become structurally significant.

As a result of the events at STP-1, on August 21, 2003, the NRC staff issued Bulletin 2003-02, "Leakage from Reactor Pressure Vessel Lower Head Penetrations and Reactor Coolant Pressure Boundary Integrity." The purpose of issuing this bulletin was to advise licensees with pressurized water reactors (PWR) units that current methods of inspecting the RPV lower heads may need to be supplemented with BMV inspections to detect reactor coolant pressure boundary (RCPB) leakage and to request licensees with PWR units to provide the NRC with information related to inspections that had been or would be performed to verify the integrity of the RPV lower head penetrations.

The staff requested this information to evaluate the integrity of the RPV lower head penetrations. The staff received the inspection plans from all 58 PWR units affected by the bulletin. A summary of these responses is listed in Section 5 of this report. The responses included the licensees' proposals to perform BMV inspections of the RPV lower head penetrations in their upcoming outage, their commitment to future inspections beyond the upcoming inspections of the RPV lower head and its penetrations, and their plans to clean the RPV lower head to establish baseline criteria for future inspections.

The NRC staff has received the inspection results from all 58 PWR units. The BMV inspections of the RPV lower head penetration were performed by 3 units during spring 2003 outage (prior to the issuance of the Bulletin 2003-02), 23 units during the fall 2003 outage, 16 units during the spring 2004 outage, 14 units during the fall 2004 outage, and 2 units during the spring 2005 outage. So far, no evidence of leakage or cracking has been found in the BMI nozzles of the RPV lower head.

The staff also issued Temporary Instruction (TI) 2515/152 - Reactor Pressure Vessel Lower Head Penetration Nozzles (NRC Bulletin 2003-02). A summary of the inspections performed by NRC regional inspectors under the TI is provided in Section 6 of this report.

The industry, which is represented by Nuclear Energy Institute (NEI), Material Reliability Project (MRP) and EPRI, is working on developing inspection and evaluation guidelines for the BMI penetrations. These guidelines are expected to address BMV inspections and may include guidelines for volumetric inspections of the BMI penetrations to monitor their aging degradation. The industry also proposed to develop new repair technology for BMI penetrations if it becomes necessary. The inspection and evaluation guidelines are not expected until 2007. A summary of these activities is provided in Section 7.

The ASME Code has developed and issued Code Case N-722, "Additional Examinations for PWR Pressure Retaining Welds in Class 1 Components Fabricated with Alloy 600/82/182 Materials ASME Section XI, Division 1," which recommends that BMV inspections be performed every other refueling outage on all the BMI penetrations in the RPV lower head. A summary of this code case is also contained in Section 7.

Section 8 of this report provides a brief summary of visual and volumetric inspections of the BMI penetrations that were performed by foreign licensees, the results of the inspections, and the future inspections plans.

Section 9 of this report provides summary and the NRC staff's conclusions.

ACKNOWLEDGMENTS

The authors sincerely thank Stephen R. Monarque for his assistance as the project manager for Bulletin 2003-02, for his valuable contributions to the development of this document, and for producing the references section of the NUREG.

The authors also thank Thomas Hiltz for his assistance in obtaining information related to the inspection results of the bottom mounted instrumentation penetrations and the associated aging monitoring programs at the foreign PWR plants.

The authors would like to express special thanks to EPRI for providing general information about the design and materials and fabrication methods of the bottom mounted instrumentation penetrations of the PWR reactor vessels.

ABBREVIATIONS

10 CFR	Title 10 to the Code of Federal Regulations	NEI	Nuclear Energy Institute
ASME	American Society of Mechanical Engineers	NRC	US Nuclear Regulatory Commission
B&W	Babcock and Wilcox	OD	Outside Diameter
BMI*	Bottom mounted instrumentation	PWHT	Post Weld Heat Treatment
BMV	Bare metal visual	PWR	Pressurized water reactor
ECT	Eddy current testing	PWSCC	Primary water stress corrosion cracking
EDM	Electrical Discharged Machine	RCPB	Reactor coolant pressure boundary
GDC	General Design Criteria	RCS	Reactor coolant system
ID	Inside diameter	RIS	Regulatory Issue Summary
ISI	Inservice inspection	RPV	Reactor Pressure Vessel
LOCA	Loss of coolant accident	STP-1	South Texas Project, Unit 1
LOF	Lack of fusion	US	United States
MRP	Materials Reliability Program	UT	Ultrasonic testing
NDE	Nondestructive examination		

*Note - In this report the following terms are used interchangeably: BMI penetrations; BMI nozzles; RPV lower head nozzles; bottom mounted nozzles; reactor vessel bottom head nozzles; reactor vessel bottom head penetrations

1 INTRODUCTION

Pressurized water reactor (PWR) reactor pressure vessel (RPV) upper heads have a number of penetrations, including upper head penetrations for the control rod drive mechanisms (CRDMs) and the lower head penetrations for nuclear in-core instrumentation. These penetrations are typically made of nickel-based Inconel Alloy 600. The penetrations are welded to the inside of the RPV head with nickel-based Inconel Alloy 82/182 materials. Cracking in PWR CRDM nozzles fabricated from Alloy 600 base material was first identified in Europe in the early 1990s. In addition, numerous small-bore Alloy 600 nozzles in the reactor coolant system (RCS) and pressurizer heater sleeves have experienced leaks; these leaks, and the cracking in CRDM nozzles has generally been attributed to primary water stress corrosion cracking (PWSCC). Most PWRs also have penetrations in the RPV lower heads for in-core nuclear instrumentation. The same Inconel materials are typically used in the lower head penetrations and welds.

The lower head and bottom mounted instrumentation (BMI) penetrations of the South Texas Project, Unit 1 (STP-1) RPV were visually inspected on April 12, 2003, as a routine part of the unit's refueling outage. The lower head of the reactor is surrounded by an insulating box structure with no insulation directly in contact with the lower head. A bare metal visual (BMV) inspection was accomplished by removing three of the insulation panels forming the insulating box. Three different vantage points were used to inspect all 58 BMI penetrations in the vessel lower head. The inspection found small amounts of white residue around two of the 58 BMI penetrations (numbers 1 and 46) at the junction where the penetrations met the lower reactor vessel head. The residue at penetrations 1 and 46 was collected for laboratory analysis to determine the source of the residue material. Approximately 150 milligrams and 3 milligrams were collected from penetrations 1 and 46, respectively. The analysis of the sample for lithium demonstrated that the lithium was approximately 99.9 percent lithium-7, which indicated that the reactor coolant system was the source of the residue. The analysis of the sample for cesium indicated that the average age of the residue collected was between 3 and 5 years. The licensee for STP-1 indicated that these residues were not visible during the previous inspection on November 20, 2002.

Ultrasonic inspections (using circumferential, axial, and zero degree probes) of 57 BMI penetration tubes at STP-1 were completed in May 2003, along with the visual inspections of the surfaces of the 58 J-groove welds which attach the BMI penetration tubes to the RPV lower head. In addition, eddy current testing (ECT) was used to examine the J-groove weld and inside diameter surfaces of some BMI penetration tubes. Axial cracks were found in penetration tubes 1 and 46. The largest of these cracks was entirely through-wall and extended above and below the J-groove weld. No evidence of cracking was found in any other penetration. BMI penetrations 1 and 46 were repaired.

The licensee performed destructive examination of one of the penetrations and determined that there was a through-wall flaw cracking due to PWSCC. The licensee's failure analysis of the two leaking penetrations determined that the root cause of the cracking was the use of Alloy 600 combined with nozzle manufacturing and installation methods that increased the susceptibility of the metal to stress corrosion cracking when in contact with primary water. PWSCC has been identified as the primary degradation mechanism affecting PWR high nickel alloy nozzles and welds (e.g., Alloy 600 tubing, piping, or forging material, and Alloy 82/182 weld material) in the RCS. PWSCC is characterized as an intergranular cracking mechanism. This mechanism occurs under conditions where a complementary combination of high welding

1

stresses, conducive environment (temperature and chemistry), and susceptible material result in premature cracking and possibly failure of the part. Even though the RPV lower head temperature is relatively low, the penetration leakage at STP-1 demonstrated that the Alloy 600 BMI nozzles are susceptible to PWSCC and will crack under the right conditions. The staff concluded that cracking at STP-1 and the small amount of leakage from the cracks did not represent an immediate safety problem due to the size and axial orientation of the cracks. However, the STP-1 experience demonstrated that BMV inspection of BMI penetrations is a useful inspection technique for detecting minor leakage and may assist in detecting flaws before they become structurally significant.

The regulations in 10 CFR 50.55a state that American Society of Mechanical Engineers (ASME) Code Class 1 components which include the reactor coolant pressure boundary (RCPB) must meet the requirements of Section XI of the ASME Code. For the RPV lower head, the ASME Code, Section XI, specifies that a qualified visual examination, called a VT-2 examination, be performed during system pressure testing. Licensees may meet the ASME Code requirement for a VT-2 inspection by performing an inspection of the RPV lower head without removing insulation from around the head and penetrations. It is the NRC staff's understanding that many licensees perform the ASME Code-required inspections without removing insulation and, therefore, may not be able to detect the amounts of through-wall leakage expected from potential flaws due to PWSCC or other cracking mechanisms. The NRC staff concluded that leakage, such as that observed at STP-1, would likely not have been detected during ASME Code inspections performed at many other PWRs.

As a result of the events at STP-1, on August 21, 2003, the NRC staff issued Bulletin 2003-02, "Leakage from Reactor Pressure Vessel Lower Head Penetrations and Reactor Coolant Pressure Boundary Integrity." The purpose of issuing this bulletin was to advise licensees with PWR units that current methods of inspecting the RPV lower heads may need to be supplemented with BMV inspections to detect RCPB leakage and to request licensees with PWR units to provide the NRC with information related to inspections that had been or would be performed to verify the integrity of the RPV lower head penetrations.

The bulletin requested that the licensees provide a description of the RPV lower head penetration BMV inspection program that will be implemented at their plants during the next and subsequent refueling outages. This inspection program was to include the plans for future inspections, the extent of the inspections, inspection methods, the process of identifying the source of findings of any boric acid deposits, and the quality of the documentation of the inspections.

The staff requested this information to evaluate the integrity of the RPV lower head penetrations. The staff received the inspection plans from all 58 PWR units affected by the bulletin. A summary of these responses is listed in Section 5 of this report. The responses included the licensees' proposals to perform BMV inspections of the RPV lower head penetrations in their upcoming outage, their commitment to future inspections beyond the upcoming inspections of the RPV lower head and its penetrations, and their plans to clean the RPV lower head to establish baseline criteria for future inspections.

The NRC staff has received the inspection results from all 58 PWR units. The BMV inspections of the RPV lower head penetration were performed by 3 units during spring 2003 outage (prior to the issuance of the Bulletin 2003-02), 23 units during the fall 2003 outage, 16 units during the spring 2004 outage,14 units during the fall 2004 outage, and 2 units during the spring 2005 outage. The staff received the inspection results from the licensees which included the type and extent of inspections, identification and characterization of boric acid deposits, and licensee's action to clean boric acid deposits from the RPV lower head to establish a baseline for future inspections. In response to an industry initiative, ultrasonic examinations have been performed at 10 units. Except for STP-1, no evidence of leakage or cracking has been found in the BMI nozzles of the RPV lower head.

2 DESCRIPTION OF VESSEL BOTTOM MOUNTED NOZZLES

The function of the bottom mounted nozzle (BMN) is to provide primary system pressure boundary-qualified entrance into the reactor pressure vessel (RPV) for the in-core instrumentation through the bottom head of the RPV. In-core instrumentation is used to monitor performance of the reactor core during operation. There are variations in the BMN designs throughout the fleet. The following provides general information about the materials and fabrication methods of the BMNs and information regarding the BMNs at South Texas Project (STP) units.

BMNs were made from Alloy 600 (trade name) SB-166, "Specification For Nickel-Chromium-Iron Alloys [Unified Numbering System (UNS) N06600, N06601, N06603, N06690, N06025 and N06045] and Nickel-Chromium-Cobalt-Molybdenum Alloy (UNS N06617) Rod, Bar And Wire," or SB-167, "Specification For Nickel-Chromium-Iron Alloys (UNS N06600, N06601, N06690, N06025, and N06045) Seamless Pipe and Tube."

BMNs were welded to the inside of the reactor vessel bottom head using partial penetration J-groove welds. The J-groove welds were made using either gas tungsten arc welding (GTAW) process with welding wire Alloy 82 (trade name) or shielded metal arc welding (SMAW) process with welding electrode Alloy 182 (trade name). Some J-groove welds were buttered, post weld heat treated (PWHT), then welded to completion without subsequent PWHT. Some J-groove welds were fully welded without butter and subsequently PWHT, depending on the particular unit and fabrication vendor. Some BMNs have circular weld pads surrounding, but not connected to, the BMNs on the outside of the reactor vessel bottom head. Such weld pads were made of either alloy steel, austenitic stainless steel, or Alloys 82/182. A comparison of Westinghouse, B&W, and CE BMN dimensions is shown in Table 2-1 of this section. A comparison of Babcock and Wilcox (B&W) and Westinghouse BMN designs is shown in Figure 2-1 of this section.

At STP Units 1 and 2 the BMNs were made of Alloy 600 material which complied with ASME SB-166 specification, and machined from 1.75" diameter bar stock. The outside diameter of the BMNs was 1.5" and the inside diameter was 0.60". The RPV material complied with ASME Specification SA-533, "Specification For Pressure Vessel Plates, Alloy Steel, Quenched And Tempered, Manganese-Molybdenum And Manganese-Molybdenum-Nickel," Grade B Class 1, with a thickness of 5.38". The RPV was cladded with 0.22" thick austenitic stainless steel weld metal. The annulus between the BMNs and the RPV lower head below the J-groove weld was 0.001" to 0.004." The J-groove welds were fabricated using shielded metal arc welding (SMAW) process with welding electrode Alloy 182. After depositing ½ of the J-groove weld, the BMNs were checked for alignment and, if required, were cold straightened. The BMNs were then welded out, ground to contour, checked for alignment, and if necessary, were cold straightened. However, the weld documentation does not provide information as to which BMN was cold straightened.

The failure analysis of the BMNs at STP Unit 1 is discussed in Section 3.0 of this document.

Manufacturer	Tube Outside Diameter	Tube Thickness	J-Groove Weld Length Parameter
Babock and Wilcox	1.03" (original) 2.0" (modified)	0.21" (original) 0.69" (modified)	1.10"
Combustion Engineering	3.0"	1.125"	1.99"
Westinghouse	1.5"	0.45"-0.587"	0.58"-1.67"

Table 2-1 Comparison of Westinghouse, Babcock and Wilcox, and Combustion Engineering Bottom Mounted Penetration Dimensions

Figure 2-1 Comparison of Westinghouse and Babcock and Wilcox Bottom
Mounted Nozzle Design

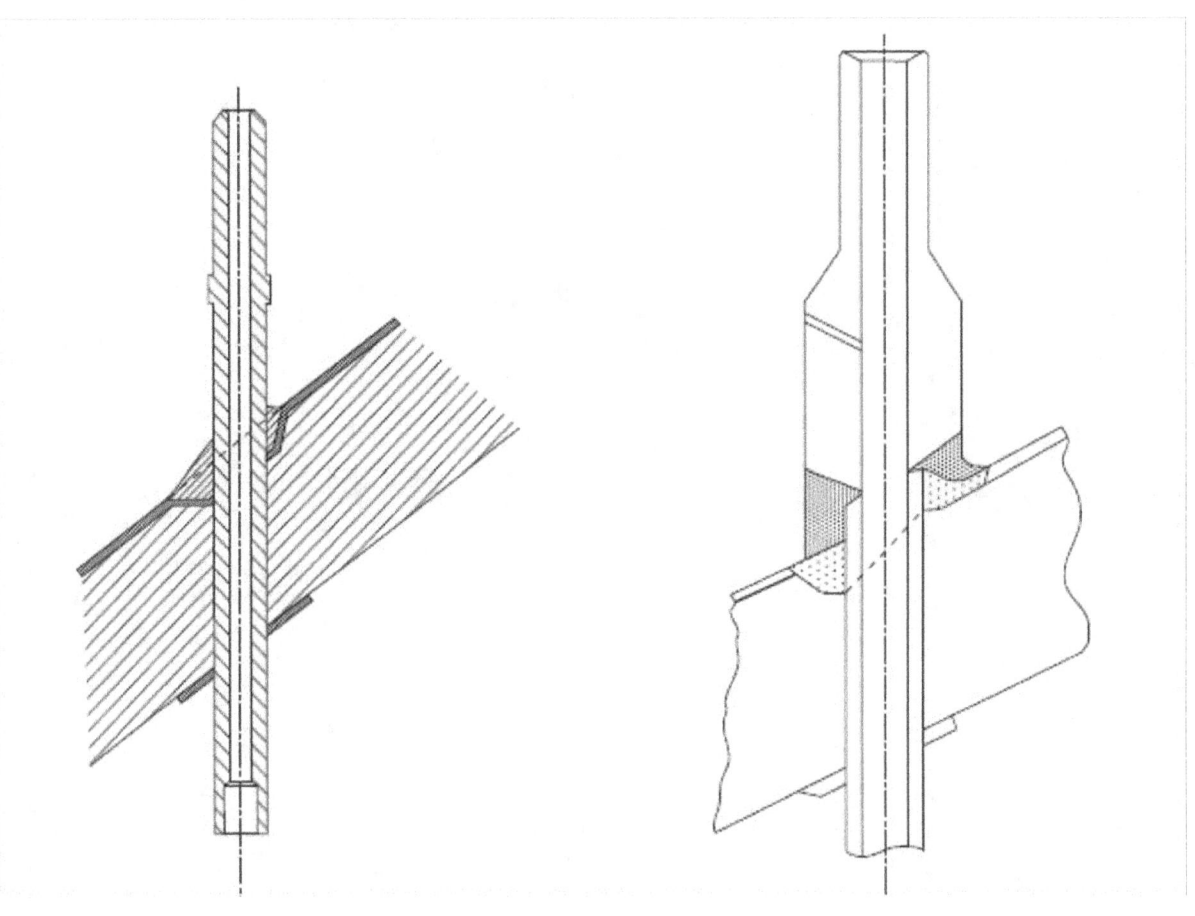

Westinghouse Design   Babcock and Wilcox Design

3 DISCUSSION OF CRACKING PHENOMENA

3.1 PRIMARY WATER STRESS CORROSION CRACKING

PWSCC has been identified as the primary degradation mechanism affecting PWR high nickel alloy nozzles and welds (e.g., Alloy 600 tubing, piping, or forging material, and Alloy 82/182 weld material). PWSCC is characterized as an intergranular cracking mechanism. This mechanism occurs under conditions where a complementary combination of high stresses (either operating or residual, typically the latter), conducive environment (temperature and chemistry), and susceptible material results in premature cracking and possibly failure of the part. High residual stresses may occur in weld materials as a result of the weld fabrication process if the welds are not stress-relieved. The shielded metal arc welding (SMAW) process used to construct the J-groove welds is prone to leaving weld defects in service and creating high residual stresses. Operating stresses resulting from reactor vessel internal pressure and/or thermal loadings may also contribute to PWSCC. An important feature of PWSCC is the crack morphology, which is characterized by very tight cracks (e.g., small crack opening displacements or angles) along material grain boundaries. These features tend to result in the potential for large through-wall flaws which may exhibit only a very small amount of leakage.

Cracking in primary side water of thicker sections was initially discovered domestically by leaks in pressurizer instrument nozzles in 1986 and in pressurizer heater sleeves in 1987. Leakage in a CRDM at a French plant was discovered in 1991. Since that time leakage or cracking has been found in other PWR components such as pressurizer relief valve nozzle safe ends, hot leg instrument nozzles, hot leg nozzle butt welds, and vessel lower head nozzles. Susceptibility to cracking has been strongly correlated to temperature and chromium content. The experience of cracking in the vessel lower head nozzles at South Texas Project, Unit 1 (STP-1) in April 2003 was not expected based on the lower temperature at this location compared to other locations in the RCS.

3.2 OPERATING EXPERIENCE

The first occurrence of cracking in a PWR BMI penetration was discovered at STP-1 when these RPV penetrations were visually inspected on April 12, 2003, as a routine part of the unit's refueling outage. The licensee was performing visual examinations of the lower vessel head during the system pressure test required by IWA-5000 of the ASME Code, Section XI. The ASME Code does not require that insulation be removed for this test. However, at STP-1 the lower head of the reactor is surrounded by an insulating box structure with no insulation directly in contact with the lower head. The inspection of the 58 BMI penetrations at this plant was a bare-metal visual inspection and was accomplished by removing three of the insulation panels forming the insulating box around and under the vessel lower head.

The inspection found small amounts of white residue around two of the 58 BMI penetrations (numbers 1 and 46) at the junction where the penetrations met the lower reactor vessel head. The residue at penetrations 1 and 46 was collected for laboratory analysis to determine the source of the residue material. Approximately 150 milligrams and 3 milligrams were collected from penetrations 1 and 46, respectively. The analysis of the sample for lithium demonstrated that the lithium was approximately 99.9 percent lithium-7, which indicated that the reactor coolant system was the source of the residue. The analysis of the sample for cesium indicated that the average age of the residue collected was between 3 and 5 years. The licensee for

STP-1 indicated that these residues were not visible during the previous bare metal visual inspection on November 20, 2002.

Ultrasonic inspections using circumferential, axial, and zero degree probes of 58 BMI penetration tubes at STP-1 were completed in May 2003, along with the visual inspections of the surfaces of the 58 J-groove welds which attach the BMI penetration tubes to the RPV lower head. In addition, eddy current testing (ECT) was used to examine the J-groove weld and inside diameter surfaces of some BMI penetration tubes. Axial cracks were found in penetration tubes 1 and 46. The largest of these cracks was entirely through-wall and extended above and below the J-groove weld.

A helium leak test was performed on the two leaking penetrations by pressurizing the annulus between the nozzle and the vessel. No bubbles were observed in Penetration 46. In Penetration 1 a small helium bubble was observed about every two seconds rising from a location outside the nozzle in the J-groove weld fillet at the tube interface.

To facilitate metallurgical analysis of the actual cracks, boat samples were removed from Penetrations 1 and 46 employing an Electric Discharge Machining (EDM) cutting technique. In the case of the BMI nozzles inside the reactor pressure vessel, the boat sample excavations could not be repaired. The desire for the largest possible boat sample was balanced against conservative structural limitations. The boat sample from Penetration 46 was designed to capture as much tube material as possible in an attempt to harvest a portion of a crack not connected to the ID of the nozzle. The margins for error associated with positioning the EDM equipment through 70 feet of water resulted in a shallow cut in Penetration 46. The resulting undersized sample was either inadvertently discarded or completely consumed in the margins of the EDM cutting tool. The boat sample from Penetration 1 captured material and defects from the J-groove weld and J-groove/tube interface, as designed. A composite drawing showing the axial crack, weld flaw and weld crack is shown in Figure 3-1.

The boat sample from Penetration 1 contained a portion of the large through-wall axial crack in the tube, three "discontinuities" which were confirmed to be lack of fusion resulting from slag inclusions, and one crack at the helium bubble location which connects the surface of the J-groove weld to the largest area of lack of fusion. The crack in the weld that connects the surface of the J-groove weld to the largest area of lack of fusion was determined from the inspections to be singular and unique. A 0.2-inch long crack spanned an 0.080-inch ligament separating the lack of fusion void from the surface of the J-groove weld in the ground fillet transition at the tube/J-groove weld interface. The length of the crack spanned and was limited to the width of the lack of fusion void. The section of the boat sample containing this crack was broken in the laboratory to expose the crack face for examination. Tenacious deposits obscured the crack face, and gradually more aggressive attempts to remove the deposits also attacked and distressed the metal surface. The crack exhibited some intergranular characteristics. To some reviewers, the nature of the oxide deposits suggested hot cracking. Fatigue could also be a factor in the development of this crack. However, the precise mechanism responsible for initiating and propagating this crack could not be determined.

Earlier ultrasonic testing (UT) results identified an axial crack in Penetration 1 which penetrated the inside diameter of the nozzle and extended from just above to just below the J-groove weld. The boat sample from Penetration 1 successfully captured a part of the upper portion of this crack in the region of the tube/J-groove weld interface. The intergranular nature of this crack exhibited classic primary water stress corrosion cracking (PWSCC) characteristics. The extent of the crack was examined by progressively grinding away thin layers of the section of the boat sample. The orientation of the ground surface was such that more weld material and less tube material was exposed at each successive grind. The initial exposed surface consisting of nearly all tube material contained a crack that extended into the weld material and then stopped. As successive layers were ground away, exposing more weld and less tube, the extent of the crack became smaller and smaller. The final ground surface, which consisted almost entirely of weld material, revealed no crack at all in the weld and a small vestige of crack in the remaining small bit of tube.

The axial crack in the tube appeared to grow from the EDM surface out to the tube/J-groove interface since it branched and connected two of the three voids, at least at this location in the boat sample. This fact might suggest inside surface initiated PWSCC. However, neither of the two cracks in Penetration 46, the other leaking BMI penetration, connected to the inside surface of the tube. A supplemental eddy current examination of the inside surface was performed, specifically to confirm the UT results that the flaws did not penetrate the inside surface. Eddy current examination established that the cracks did not connect to the inside surface. Based on this fact the licensee concluded that the PWSCC axial crack in the tube was OD initiated. The crack most likely originated on the outside surface of the tube in the highly stressed region of the flooded weld defects.

The boat sample also contained numerous small cracks around the periphery of the tube containing lack of fusion (LOF) voids to a depth of 1 or 2 grains. Although hot cracking in the weld material is a possibility, this intergranular cracking also appeared in the nozzle, where hot cracking is not possible. Therefore, the licensee concluded that this cracking is PWSCC resulting from flooding of the LOF voids.

In summary, metallurgical analysis of a sample removed from one of the leaking BMI penetrations confirmed the presence of weld defects on the highly stressed interface between the Alloy 600 tube and the connection weld to the pressure vessel. The SMAW process used to construct the J-groove welds is prone to leaving weld defects in service and creating locally high residual stresses. The sample revealed one small crack that connected the LOF void to the surface of the weld and primary water. Once the LOF void became flooded with primary water, all of the requisite conditions to support stress corrosion cracking existed at the nozzle outside surface at a location of predicted high residual stress.

The penetration leakage at South Texas demonstrated that the Alloy 600 BMI nozzles are susceptible to PWSCC and will crack under the right conditions. Even at the lower temperatures of the bottom head, PWSCC is possible. Additionally, the shielded metal arc welding (SMAW) process used to construct the J-groove welds is prone to leaving weld defects in service and creating high residual stresses. The licensee did not identify any materials or fabrication techniques unique to the construction of the STP-1 reactor vessel related to the occurrence of these cracks.

The STP-1 experience also demonstrates that visual examination of bare metal BMI penetrations is an effective mechanism for detecting minor leakage. The root cause is the use of Alloy 600 combined with nozzle manufacturing and installation methods that further increased the susceptibility of the metal to stress corrosion cracking when in contact with primary water.

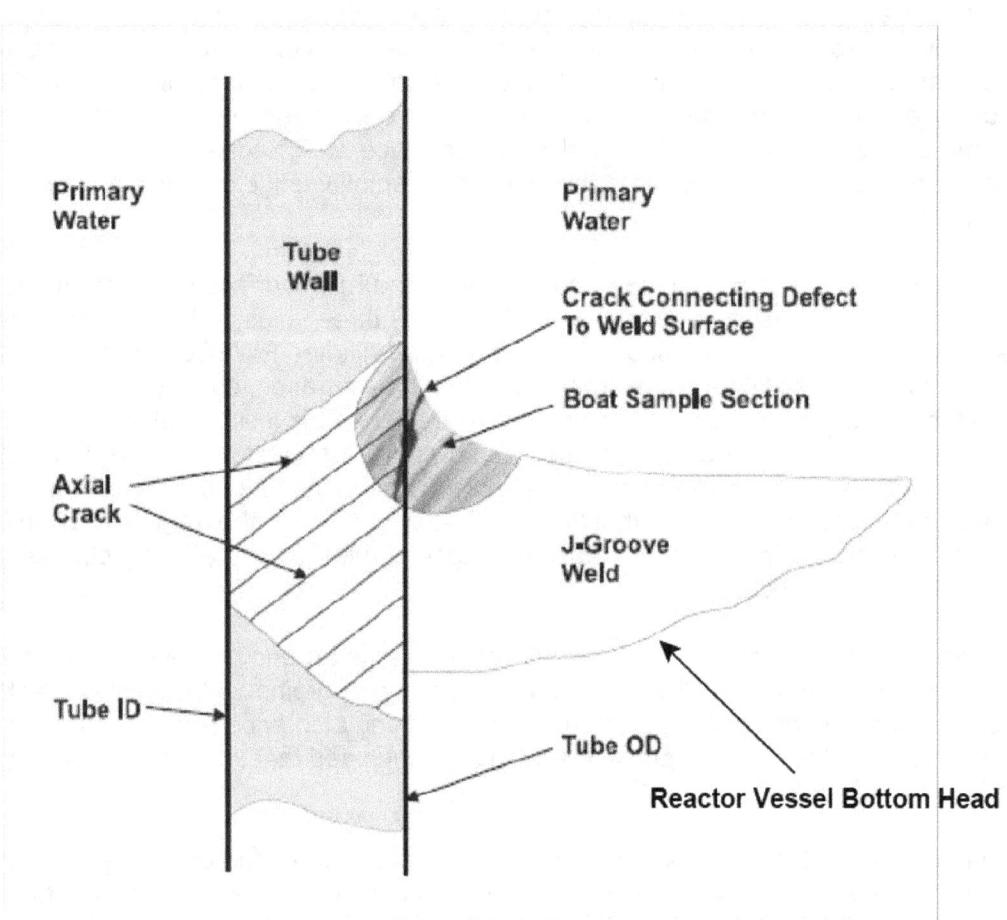

Figure 3-1 Composite Drawing Showing Axial Crack, Weld Flaw, and Weld Crack

4 DESCRIPTION OF BULLETIN 2003-02

4.1 SUMMARY OF BULLETIN

The bulletin was addressed to all holders of operating licenses for pressurized-water nuclear power reactors (PWRs) with penetrations in the lower head of the reactor pressure vessel (RPV). The addressees included PWRs designed by Westinghouse and Babcock and Wilcox, but did not include PWRs designed by Combustion Engineering, except for the Palo Verde units. All subject PWR addressees were requested to provide the following information within 30 days of the date of the bulletin if the facility would be entering a refueling outage before December 31, 2003, or within 90 days otherwise.

(1)   A description of the RPV lower head penetration inspection program that has been implemented at the plant. The description should include when the inspections were performed, the extent of the inspections with respect to the areas and penetrations inspected, inspection methods used, the process used to resolve the source of findings of any boric acid deposits, the quality of the documentation of the inspections (e.g., written report, video record, photographs), and the basis for concluding that the plant satisfies applicable regulatory requirements related to the integrity of the RPV lower head penetrations.

(2)   A description of the RPV lower head penetration inspection program that will be implemented at the plant during the next and subsequent refueling outages. The description should include the extent of the inspections which will be conducted with respect to the areas and penetrations to be inspected, inspection methods to be used, qualification standards for the inspection methods, the process used to resolve the source of findings of boric acid deposits or corrosion, the inspection documentation to be generated, and the basis for concluding that the plant will satisfy applicable regulatory requirements related to the structural and leakage integrity of the RPV lower head penetrations.

Within 60 days of plant restart following the next inspection of the RPV lower head penetrations, the subject PWR addressees were requested to submit to the NRC a summary of the inspections performed, the extent of the inspections, the methods used, a description of the as-found condition of the lower head, any findings of relevant indications of through-wall leakage, and a summary of the disposition of any findings of boric acid deposits and any corrective actions taken as a result of indications found.

4.2 SUMMARY OF REGULATORY ISSUES

4.2.1 APPLICABLE REGULATORY REQUIREMENTS

Several provisions of the NRC regulations and plant operating licenses (technical specifications) pertain to RCPB integrity and the issues addressed by Bulletin 2003-02. The general design criteria (GDC) for nuclear power plants (Appendix A to 10 CFR Part 50), or, as appropriate, similar requirements in the licensing basis for a reactor facility, the requirements of 10 CFR 50.55a pertaining to the ASME Code, and the quality assurance criteria of Appendix B to 10 CFR Part 50 provide the bases and requirements for NRC staff assessment of the potential for, and consequences of, degradation of the RCPB.

The applicable GDCs include GDC 14 (Reactor Coolant Pressure Boundary), GDC 31 (Fracture Prevention of Reactor Coolant Pressure Boundary), and GDC 32 (Inspection of Reactor Coolant Pressure Boundary). GDC 14 specifies that the RCPB be designed, fabricated, erected, and tested so as to have an extremely low probability of abnormal leakage, of rapidly propagating failure, and of gross rupture. GDC 31 specifies that the probability of rapidly propagating fracture of the RCPB be minimized. GDC 32 specifies that components which are part of the RCPB have the capability of being periodically inspected to assess their structural and leaktight integrity.

NRC regulations in 10 CFR 50.55a state that ASME Class 1 components (which includes the RCPB) must meet the requirements of Section XI of the ASME Code. Various portions of the ASME Code address RCPB inspection. For example, Table IWB-2500-1 of Section XI of the ASME Code provides examination requirements during system leakage testing of all pressure-retaining components of the RCPB and references IWB-3522 for acceptance standards. IWB-3522.1(c) and (e) specify that conditions requiring correction include the detection of leakage from insulated components and discoloration or accumulated residues on the surfaces of components, insulation, or floor areas that may be evidence of borated water leakage, with leakage defined as the through-wall leakage that penetrates the pressure retaining membrane. Therefore, 10 CFR 50.55a, by reference to the ASME Code, does not permit through-wall degradation of the RPV lower head penetrations. For through-wall leakage identified by visual examinations in accordance with the ASME Code, acceptance standards for the identified degradation are provided in IWB-3142. Specifically, supplemental examination (by surface or volumetric examination), corrective measures or repairs, analytical evaluation, and replacement provide methods for determining the acceptability of degraded components.

Criterion V (Instructions, Procedures, and Drawings) of Appendix B to 10 CFR Part 50 states that activities affecting quality shall be prescribed by documented instructions, procedures, or drawings of a type appropriate to the circumstances and shall be accomplished in accordance with these instructions, procedures, or drawings. Criterion V further states that instructions, procedures, or drawings shall include appropriate quantitative or qualitative acceptance criteria for determining that important activities have been satisfactorily accomplished. Visual and volumetric examinations of the RCPB are activities that should be documented in accordance with these requirements.

Criterion IX (Control of Special Processes) of Appendix B to 10 CFR Part 50 states that special processes, including nondestructive testing, shall be controlled and accomplished by qualified personnel using qualified procedures in accordance with applicable codes, standards, specifications, criteria, and other special requirements.

Criterion XVI (Corrective Action) of Appendix B to 10 CFR Part 50 states that measures shall be established to assure that conditions adverse to quality are promptly identified and corrected. For significant conditions adverse to quality, the measures taken shall include root cause determination and corrective action to preclude repetition of the adverse conditions. For degradation of the RCPB, the root cause determination is important for understanding the nature of the degradation present and the required actions to mitigate future degradation. These actions could include proactive inspections and repair of degraded portions of the RCPB.

Plant technical specifications (TS) pertain to this issue insofar as they do not allow operation with through-wall reactor coolant system pressure boundary leakage.

4.2.2 CURRENT REQUIREMENTS AND THE REASON FOR ISSUING BULLETIN 2003-02

The NRC issued Regulatory Issue Summary (RIS) 2003-13, "NRC Review of Responses to Bulletin 2002-01, "Reactor Pressure Vessel Head Degradation and Reactor Coolant Pressure Boundary Integrity," on July 29, 2003. This RIS was issued based on the results of the staff's review of the responses to Bulletin 2002-01 on reactor coolant system boric acid corrosion. The NRC noted in RIS 2003-13 that most licensees do not perform inspections of Alloy 600/82/182 materials beyond those required by Section XI of the ASME Code to identify potential cracked and leaking components. For the RPV lower head, the ASME Code specifies that a visual examination, called a VT-2 examination, be performed during system pressure testing. Licensees may meet the ASME Code requirement for a VT-2 inspection by performing an inspection of the RPV lower head without removing insulation from around the head and penetrations. It is the NRC staff's understanding that many licensees perform the ASME Code-required inspections without removing insulation and, therefore, may not be able to detect the amounts of through-wall leakage expected from potential flaws due to PWSCC or other cracking mechanisms. From the NRC staff reviews described in RIS 2003-13, the NRC staff concluded that leakage such as that observed at STP-1 would likely not have been detected during ASME Code inspections performed at many other PWRs.

The circumstances of the STP-1 findings indicate that the cracking and the onset of leakage may have occurred several years prior to the discovery of leakage. This licensee's prior inspections of STP-1 lower head were capable of finding the deposits observed in April 2003. However, no evidence of leakage had been noted as the result of any inspections conducted prior to April 2003. Therefore, the staff concluded that inspections of the RPV lower head area beyond those required by the ASME Code were and continue to be appropriate for ensuring that there are not leaks from the lower head penetrations. Therefore, Bulletin 2003-02 was issued requesting the information summarized above and stated that inspections capable of detecting through-wall leakage from any RPV lower head penetration, beginning at the next refueling outage, would provide additional confidence in the integrity of the RPV lower head penetrations.

The NRC staff is working with the industry and other stakeholders to revise the ASME Code to address inspection of RCPB locations susceptible to cracking, including RPV penetrations. These activities will not be completed for several years, so the NRC issued Bulletin 2003-02 to address the immediate concerns discuss above.

5 RESPONSE FROM PLANTS

The US Nuclear Regulatory Commission (NRC) issued the bulletin 2003-02 to:

(1)   advise PWR licensees that current methods of inspecting the reactor pressure vessel (RPV) lower heads may need to be supplemented with additional measures (e.g., bare-metal visual inspections) to detect reactor coolant pressure boundary (RCPB) leakage and

(2)   request PWR licensees to provide the NRC with information related to inspections that have been or will be performed to verify the integrity of the RPV lower head penetrations.

The bulletin stated that it is appropriate for licensees to assess their current inspection practices to periodically ensure that there are no leaks from RPV lower head penetrations. This conclusion was based on the safety concerns associated with a significant leak from the RPV lower head and the uncertainties associated with the ability of some current inspection practices to identify cracks and resultant small leaks from RPV lower head penetrations.

5.1 DESCRIPTION OF RESPONSES REQUESTED

The staff requested all subject PWR licensees to provide the following information. The responses for facilities entering refueling outages before December 31, 2003, were requested to provide responses within 30 days of the date of the bulletin. All other responses were requested to be provided within 90 days of the date of the bulletin:

(1)   The licensee were asked to provide a description of the RPV lower head penetration BMV inspection program that will be implemented at the plant during the next and subsequent refueling outages. The inspection program was requested to include the plans for future inspections, the extent of the inspections, inspection methods, the process of identifying the source of findings of any boric acid deposits, and the quality of the documentation of the inspections.

(2)   If the licensee did not plan to perform either a bare-metal visual inspection at the next or subsequent refueling outages, it was requested to provide a basis for concluding that the inspections performed thus far, would assure applicable regulatory requirements were and would continue to be met.

Within 60 days of plant restart following the next inspection of the RPV lower head penetrations, the subject PWR licensees were requested to submit to the NRC a summary of the inspections performed, the extent of the inspections, the methods used, a description of the as-found condition of the lower head, any findings of relevant indications of through-wall leakage, and a summary of the disposition of any findings of boric acid deposits and any corrective actions taken as a result of indications found.

5.2 INITIAL RESPONSES

5.2.1 SUMMARY OF INFORMATION ON TABLE OF INITIAL RESPONSES

The licensees were requested to provide the staff with information related to current inspections and future inspections of the RPV lower head penetrations. The staff received the initial responses from all 58 units, and a summary of these responses is listed in Table 5.2.1 of this section. The table is based on the following attributes from each response.

(1) the licensee's proposal to perform bare metal visual (BMV) inspections of the RPV lower head penetrations in upcoming outage,

(2) the licensee's disposition findings of deposits at the RPV lower head and its penetrations,

(3) the licensee's commitment to future inspections beyond the upcoming inspections of the RPV lower head and its penetrations, and

(4) the licensee's plans to clean the RPV lower head and establish baseline criteria for future inspections.

A listing of the initial responses can be found in references section in Appendix B to this NUREG. The references section provides information related to the plant name and accession number of each submittal.

Inspection Attributes\Plant	Arkansas Nuclear, Unit 1	Beaver Valley, Units 1 and 2	Braidwood Station, Units 1 and 2	Byron Station, Units 1 and 2
(1) Proposal to perform bare metal visual (BMV) inspection in upcoming outage	Spring 2004 outage - BMV inspection of each penetration using direct or equivalent remote visual aided by remote camera.	Fall 2003 outage Unit 2; Fall 2004 outage Unit 1 - BMV inspection of each penetration using direct or equivalent remote visual aided by remote camera.	Fall 2003 outage Unit 2; Fall 2004 outage Unit 1 - BMV inspection of each penetration using a robotic camera.	Fall 2003 outage Unit 1; Spring 2004 outage Unit 2 - BMV inspection of each penetration using a robotic camera.
(2) Plans to disposition findings of deposits	Chemical analysis will be done of any suspicious looking deposit with discernable thickness.	Chemical analysis will be done of any suspicious looking deposit with discernable thickness.	Chemical analysis will be done of any suspicious looking deposit with discernable thickness.	Chemical analysis will be done of any suspicious looking deposit with discernable thickness.
(3) Commitment to future inspections beyond the upcoming inspections	Scope and approach for future nozzle inspections (beyond Spring 2004) will consider lessons learned from initial inspection and other BMI inspections conducted in the industry.	BMV of each penetration will be performed at each refueling outage until ASME Code changes or regulatory action justifies a change in the examination frequency or method.	Inspections during the subsequent refueling outages will be performed based on inspection findings, industry developments and NRC guidance.	Inspections during the subsequent refueling outages will be performed based on inspection findings, industry developments and NRC guidance.

Table 5.2.1 Summary of Initial Bulletin Responses

19

Inspection Attributes\Plant	Callaway, Unit 1	Catawba, Units 1 and 2	Comanche Peak, Units 1 and 2	Crystal River, Unit 3
(1) Proposal to perform BMV inspection in upcoming outage	Spring 2004 - all 58 BMI penetrations, including 100% of circumference of each penetration. Direct visual VT-2; certified VT-2 level II or III.	Fall 2003 outage Unit 1; Fall 2004 outage Unit 2 - BMV inspection of each penetration 360^0 using robotic camera.	Fall 2003 outage Unit 2; Spring 2004 outage Unit 1 - BMV inspection of each penetration using a robotic camera.	Fall 2003 outage - BMV inspection of each penetration will be performed. Method of inspection was not specified.
(2) Plans to disposition findings of deposits	Chemical analysis will be done of any suspicious looking deposit with discernable thickness.	Chemical analysis will be done of any suspicious looking deposit with discernable thickness.	Chemical analysis will be done of any suspicious looking deposit with discernable thickness.	Chemical analysis will be done of any suspicious looking deposit with discernable thickness.
(3) Commitment to future inspections beyond the upcoming inspections	Will perform bare metal visual inspections at subsequent outages.	Inspections during the subsequent refueling outages will be performed based on inspection findings and industry developments.	Inspections will be performed every five years or every third refueling, whichever occurs first until industry experience provides sound basis for a change in inspection frequency or method.	Enhanced visual inspection of all the BMI will be performed during 2007 outage in conjunction with 10 year ISI.

Table 5.2.1 Summary of Initial Bulletin Responses

Inspection Attributes\Plant	D.C. Cook, Units 1 and 2	Davis-Besse	Diablo Canyon, Units 1 and 2	Joseph M. Farley, Units 1 and 2
(1) Proposal to perform BMV inspection in upcoming outage	Fall 2003 outage Unit 1; Fall 2004 outage Unit 2 BMV inspection of each penetration.	Spring 2003 outage BMV inspection of each penetration using direct or equivalent remote visual aided by remote camera.	Spring 2004 outage Unit 1, Fall 2004 outage Unit 2 examine all 58 BMI penetrations, 100% of circumference, by direct or remote visual.	Spring 2003 outage Unit 1; Spring 2004 outage Unit 2 - BMV inspection of each penetration using a robotic camera.
(2) Plans to disposition findings of deposits	Chemical analysis will be done of any suspicious looking deposit with discernable thickness.	Chemical analysis will be done of any suspicious looking deposit with discernable thickness.	May perform sampling and isotopic analysis.	Chemical analysis will be done of any suspicious looking deposit with discernable thickness.
(3) Commitment to future inspections beyond the upcoming inspections	Inspections during the subsequent refueling outages will be performed for each unit.	BMV inspection would apply to next outage. No additional commitments were made for future outages.	Repeat visual examination of all 58 BMI penetrations every third outage.	No commitment was made regarding BMV inspections during subsequent outages. See report summary and conclusions, Section 9.

Table 5.2.1 Summary of Initial Bulletin Responses

Inspection Attributes\Plant	Ginna	Indian Point, Units 2 and 3	Kewaunee	McGuire, Units 1 and 2
(1) Proposal to perform BMV inspection in upcoming outage	Fall 2003 outage - BMV inspection of each penetration with a camera on a pole or other video device.	Fall 2004 outage Unit 2; Spring 2005 outage Unit 3 - BMV inspection of each penetration using direct or equivalent remote visual aided by remote camera.	Fall 2004 outage- BMV inspection of each penetration.	Fall 2003 outage Unit 2; Spring 2004 outage Unit 1 - BMV inspection of 360^0 of each penetration using robotic camera.
(2) Plans to disposition findings of deposits	Chemical analysis will be done of any suspicious looking deposit with discernable thickness.	Chemical analysis will be done of any suspicious looking deposit with discernable thickness.	Chemical analysis will be done of any suspicious looking deposit with discernable thickness.	Chemical analysis will be done of any suspicious looking deposit with discernable thickness.
(3) Commitment to future inspections beyond the upcoming inspections	BMV of lower head penetrations will be performed during each refueling outage until changes to the ASME Code or industry recommendations justify a change in the examination frequency.	BMV inspection would apply to subsequent outages, unless industry experience or site-specific observations indicate the need for an alternate inspection approach.	BMV inspection would apply to subsequent outages until changes to the ASME Code or industry recommendations justify a change in the examination frequency.	Inspections during the subsequent refueling outages will be performed based on inspection findings and industry developments.

Table 5.2.1 Summary of Initial Bulletin Responses

Inspection Attributes\Plant	Millstone, Unit 3	North Anna, Units 1 and 2	Oconee, Units 1, 2 and 3	Palo Verde, Units 1, 2, and 3
(1) Proposal to perform BMV inspection in upcoming outage	Spring 2004 outage - BMV inspection of 360^0 of each penetration using direct or equivalent remote visual aided by remote camera.	Spring 2004 outage Unit 2; Fall 2004 outage Unit 1 - BMV inspection of 360^0 of each penetration using direct or equivalent remote visual aided by remote camera.	Fall 2003 outage Unit 1; Spring 2004 outage Unit 2 and Fall 2004 outage Unit 3 - BMV inspection of 360^0 of each penetration using robotic camera.	Fall 2003 outage Unit 2; Spring 2004 outage Unit 1; and Fall 2004 Unit 3, Spring 2005 outage Unit 2 - BMV inspection of each penetration using a robotic camera.
(2) Plans to disposition findings of deposits	Chemical analysis will be done of any suspicious looking deposit with discernable thickness.	Chemical analysis will be done of any suspicious looking deposit with discernable thickness.	Chemical analysis will be done of any suspicious looking deposit with discernable thickness.	The process includes an evaluation to determine if leakage has occurred and identify the source of leakage.
(3) Commitment to future inspections beyond the upcoming inspections	Inspections during the subsequent refueling outages will be performed based on inspection findings and industry developments.	Inspections during the subsequent refueling outages will be performed based on inspection findings and industry developments.	Inspections during the subsequent refueling outages will be performed based on inspection findings and industry developments.	Inspections during the subsequent refueling outages will be performed based on inspection findings and industry developments and guidance set up by the NRC staff.

Table 5.2.1 Summary of Initial Bulletin Responses

Inspection Attributes\Plant	Point Beach, Units 1 and 2	Prairie Island, Units 1 and 2	H.B. Robinson, Unit 2	Salem, Units 1 and 2
(1) Proposal to perform BMV inspection in upcoming outage	Fall 2003 outage Unit 2; Spring 2004 outage Unit 1 - BMV inspection of each penetration using a robotic camera.	Fall 2003 outage Unit 2 - BMV inspection of each penetration using a robotic camera.	Spring 2004 outage - BMV inspection of each penetration using a robotic camera.	Fall 2003 outage Unit 2; Spring 2004 outage Unit 1, - BMV inspection of 360^0 of each penetration using a robotic camera.
(2) Plans to disposition findings of deposits	Chemical analysis will be done of any suspicious looking deposit with discernable thickness.	Chemical analysis will be done of any suspicious looking deposit with discernable thickness.	Evaluation of any suspicious looking deposit is performed in accordance with boric acid corrosion control program and technical specification.	No details are provided for verification of boric acid deposits.
(3) Commitment to future inspections beyond the upcoming inspections	Inspections during the subsequent refueling outages will be performed for each unit.	Inspections during the subsequent refueling outages will be performed for each unit.	The periodicity and the scope of the future inspections will be based on the results of the inspections and regulatory guidance.	Inspections during the subsequent refueling outages will be performed.

Table 5.2.1 Summary of Initial Bulletin Responses

24

Inspection Attributes\Plant	Seabrook	Sequoyah, Units 1 and 2	Shearon Harris, Unit 1	South Texas, Units 1 and 2
(1) Proposal to perform BMV inspection in upcoming outage	Fall 2003 outage - BMV inspection of each penetration using a robotic camera.	Fall 2003 outage Unit 2; Spring 2003 Unit 1- BMV inspection of each penetration using a robotic camera.	Fall 2004 outage - BMV inspection of each penetration using direct or equivalent remote visual aided by remote camera.	Unit 2 Fall 2005 - BMV inspection of each penetration using direct or equivalent remote visual aided by remote camera.
(2) Plans to disposition findings of deposits	Chemical analysis will be done of any suspicious looking deposit with discernable thickness.	Chemical analysis will be done of any suspicious looking deposit with discernable thickness.	Chemical analysis will be done of any suspicious looking deposit with discernable thickness.	Unit 1 verified deposits due to RCS leakage. Plans to disposition any Unit 2 boric acid deposits not addressed.
(3) Commitment to future inspections beyond the upcoming inspections	BMV inspections during the subsequent refueling outages will be performed for each unit until ASME Code changes or regulatory actions justify a change in this frequency.	BMV inspections during the subsequent refueling outages will be performed for each unit until ASME Code changes or regulatory actions justify a change in this frequency.	Not addressed. See report summary and conclusions, Section 9.	BMV inspections during the subsequent refueling outages will be performed for each unit.

Table 5.2.1 Summary of Initial Bulletin Responses

Inspection Attributes\Plant	V.C. Summer	Surry, Units 1 and 2	Three Mile Island, Unit 1	Turkey Point, Units 3 and 4
(1) Proposal to perform BMV inspection in upcoming outage	Fall 2003 outage BMV inspection of each penetration using a robotic camera will be performed.	Fall 2003 outage Unit 2; Fall 2004 outage Unit 1 - BMV inspection 360^{0} of each penetration using a robotic camera.	Fall 2003 outage BMV inspection of each penetration using a robotic camera.	Fall 2003 outage Unit 4; Fall 2004 outage Unit 3- BMV inspection of each penetration using a robotic camera.
(2) Plans to disposition findings of deposits	Chemical analysis will be done of any suspicious looking deposit with discernable thickness.	Chemical analysis will be done of any suspicious looking deposit with discernable thickness.	Chemical analysis will be done of any suspicious looking deposit with discernable thickness.	Chemical analysis will be done of any suspicious looking deposit with discernable thickness.
(3) Commitment to future inspections beyond the upcoming inspections	The scope of BMV inspections during the subsequent refueling outages will be dependent on industry guidance.	Inspections during the subsequent refueling outages will be performed. Less frequent inspections may be adapted based on industry's ongoing research.	Inspections during the subsequent refueling outages will be performed based on inspection findings, industry developments, and NRC guidance.	Inspections during the subsequent refueling outages will be performed based on inspection findings, industry developments, and NRC guidance.

Table 5.2.1 Summary of Initial Bulletin Responses

Inspection Attributes\Plant	Vogtle, Units 1 and 2	Watts Bar, Unit 1	Wolf Creek
(1) Licensee's proposal to perform BMV inspection in upcoming outage	Fall 2003 outage Unit 1; Spring 2004 outage Unit 2 - BMV inspection of each penetration using a robotic camera.	Fall 2003 outage - BMV inspection of each penetration using a robotic camera will be performed.	Fall 2003 outage - BMV inspection of each penetration without a robotic camera.
(2) Plans to disposition findings of deposits	Chemical analysis will be done of any suspicious looking deposit with discernable thickness.	Chemical analysis will be done of any suspicious looking deposit with discernable thickness.	Chemical analysis will be done of any suspicious looking deposit with discernable thickness.
(3) Commitment to future inspections beyond the upcoming inspections	No commitment on future BMV inspections was made. See report summary and conclusions, Section 9.	BMV inspections during the subsequent refueling outages will be performed until ASME Code changes or regulatory actions justify a change in this frequency.	No commitment was made regrading BMV inspection of the bottom head during the subsequent outages. See report summary and conclusions, Section 9.

Table 5.2.1 Summary of Initial Bulletin Responses

27

5.3 RESPONSES ON INSPECTION RESULTS

5.3.1 SUMMARY OF INFORMATION ON TABLE OF INSPECTION RESULTS

PWR licensees were requested to submit to the NRC a summary of the inspections performed, the extent of the inspections, the methods used, a description of the as-found condition of the lower head, any findings of relevant indications of through-wall leakage, and a summary of the disposition of any findings of boric acid deposits and any corrective actions taken as a result of indications found.

The NRC staff has received the inspection results from all 58 PWR units. The BMV inspections of the RPV lower head penetration were performed by 3 units during spring 2003 outage (prior to the issuance of the Bulletin 2003-02), 23 units during the fall 2003 outage,16 units during the spring 2004 outage, and 14 units during the fall 2004 outage. During the spring 2005, BMV inspections of the BMI penetrations were performed on 2 units. In addition, during the spring 2005 outage, the licensee for Palo Verde Unit 2 conducted a follow-up inspections on the BMI penetrations which is discussed below. The inspection results of each unit are summarized in Tables 5.3-1, 5.3-2, 5.3-3, 5.3-4 and 5.3-5 of this NUREG. Each table contains the results of the inspections that took place during the five refueling outage seasons between the spring of 2003 and the spring of 2005. A summary of the inspection results is provided in tables below and the summary addresses the following attributes from each licensee's response.

(1) Type and extent of inspection for example, inspection with robotic camera or pole mounted camera; 360° coverage of each nozzle.

(2) Identification of boric acid deposits and characterization of deposits.

(3) Chemical analysis, number of samples of the deposit, and results.

(4) Licensee's actions during outage to clean boric acid deposits from the RPV lower head and to establish a baseline for future inspections.

The licensees for the following units conducted BMV inspections on their RPV lower head penetrations during the Spring 2003 outage prior to the issuance of the Bulletin 2003-02 and these inspection results are summarized in Table 5.3-1.

(1) Davis-Besse; (2) Joseph M. Farley, Unit 1; (3) Sequoyah Unit 1.

A listing of the responses containing the inspection results can be found in the references section of Appendix B to this NUREG. The references section provides information related to the plant name and accession number of each submittal.

The licensee for the Palo Verde, Unit 2 conducted its first BMV inspections of the RPV lower head penetrations during the fall 2003 outage. During the inspection it was discovered that the annulus area between the penetrations and the RPV lower head was covered with a corrosion protective coating of spraylat (trade name). To achieve a proper inspection of each BMI penetration, the licensee cleaned the annulus area between the penetrations and the RPV lower head. However, due to the equipment problems it could only complete the cleaning of 39 out of 61 penetrations. The inspection results of the 39 penetrations are summarized in Table 5.3-2. During the spring 2005 outage, the licensee cleaned and inspected the remaining 22 penetrations of the RPV lower head at the Palo Verde Unit 2. The inspection results of the 22 penetrations are summarized in Table 5.3-5.

5.3.2 TABLE 5.3-1, INSPECTION RESULTS, SPRING 2003 OUTAGE

Inspections performed prior to the issuance of the Bulletin 2003-02

Inspection Attributes\Plant	Davis Besse Nuclear Power Station
(1) Type and extent of inspection	All BMI penetrations were examined 360^0 around each circumference with a video camera by VT-2 qualified inspectors.
(2) Identification of boric acid deposits and characterization of deposits	The licensee identified stains with no discernable thickness at some of the RPV lower head penetrations. Based on the appearance and texture of these stains, the licensee stated that there was no evidence of boric acid leakage at the RPV lower head penetrations.
(3) Chemical analysis, number of samples of the deposit, and results	The licensee confirmed the absence of RCS leakage by performing chemical analysis of the stained deposits. Based on the results of the chemical analysis, the licensee concluded that there was no RCS leakage.
(4) Licensee's actions during outage to clean boric acid deposits from the vessel bottom head to establish a baseline for future inspections	The licensee cleaned the RPV lower head and performed BMV of the RPV lower head penetrations and no boric acid deposits were found at these locations.

Table 5.3-1, Inspection Results, Spring 2003 Outage

Inspections performed prior to the issuance of the Bulletin 2003-02

Inspection Attributes\Plant	Joseph M. Farley, Unit 1
(1) Type and extent of inspection	All 50 BMI penetrations were examined 360^0 around each circumference by VT-2 qualified inspectors.
(2) Identification of boric acid deposits and characterization of deposits	The licensee identified translucent white residue streams on the RPV lower head. The licensee observed no boric acid deposits of discernable thickness on the vessel lower head. The rust trails appeared to have originated from above the vessel lower head. Thus, the licensee concluded that there was no leakage at the lower head penetrations.
(3) Chemical analysis, number of samples of the deposit and results	The licensee did not perform chemical analysis.
(4) Licensee's actions during outage to clean boric acid deposits from the vessel lower head to establish a baseline for future inspections	Not addressed.

Table 5.3-1, Inspection Results, Spring 2003 Outage

Inspections performed prior to the issuance of the Bulletin 2003-02

Inspection Attributes\Plant	Sequoyah, Unit 1
(1) Type and extent of inspection	All 58 BMI penetrations on the RPV lower head were examined around each circumference with a video camera attached to a magnetic crawler device and the inspection results were recorded on VHS videotape. Visual examination of the RPV lower head area was performed by VT-2 inspectors.
(2) Identification of boric acid deposits and characterization of deposits	The licensee identified white residue streams with no discernable thickness on the RPV lower head. The licensee did not observe any white boric acid residue on the RPV lower head. Thus, the licensee concluded that there was no leakage at the lower head penetrations.
(3) Chemical analysis, number of samples of the deposit, and results	The licensee did not perform chemical analysis.
(4) Licensee's actions during outage to clean boric acid deposits from the vessel bottom head to establish a baseline for future inspections	Not addressed.

Table 5.3-1, Inspection Results, Spring 2003 Outage

Inspection Attributes\Plant	Beaver Valley, Unit 2
(1) Type and extent of inspection	Visual inspection of 50 bottom mounted instrument (BMI) penetrations including 100% of the circumference of each penetration annulus was done using a remote crawler with a zoom camera. Each penetration was divided into four quadrants, and still images and video tape were made for each penetration. Each penetration's identity was indexed. Inspection was performed by VT-2 inspectors.
(2) Identification of boric acid deposits and characterization of deposits	Presence of debris was noted on eighteen penetrations, and on each boss surface around the each penetration. The debris appeared to be from protective Spraylat (trade name) coating and from tape. Milky white streak deposits were noted on the insulation which was attributed to previous reactor cavity leaks.
(3) Chemical analysis, number of samples of the deposit and results	Chemical analysis from the vicinity of annulus showed concentration of less than 0.1 ppm (lowest detection limit) of lithium and boron. Three additional samples were taken from the tape to confirm the presence of halogens and heavy metals. Only the presence of sulphur and other elements that are typically present in the protective tape were detected. All these elements were determined to be non-detrimental. Based on these results, the licensee concluded that there was no RCS leakage.
(4) Licensee's actions during outage to clean boric acid deposits from the vessel bottom head to establish a baseline for future inspections	Licensee will clean the residue at the annulus region of the penetration during the next spring 2005 outage. Fall 2003 visual inspection data (video tape) will be used as a baseline for the spring 2005 outage for Unit 2.

Table 5.3-2, Inspection Results, Fall 2003 Outage

Inspection Attributes\Plant	Braidwood Station, Unit 2
(1) Type and extent of inspection	All 58 penetrations were examined 360^0 around by certified VT examiners, using a remotely operated zoom lens camera.
(2) Identification of boric acid deposits and characterization of deposits	The licensee observed boric acid accumulation at the annulus region of the penetration 45. The licensee postulated that this accumulation originated from the cavity seal leakage that occurred in spring 1996.
(3) Chemical analysis, number of samples of the deposit and results	Radiological smear of penetration 45 showed no signs of RCS activity and only naturally occurring nuclides were present. There are a number of possible sources for these radioactive species. The licensee did not or was not able to collect and analyze samples for the presence of boron and lithium. Therefore, the chemical analysis was of limited value. However, the licensee observed no boric acid deposits of discernable thickness on the bottom head. On this basis of observation, the licensee concluded that there was no leakage at the RPV lower head penetrations.
(4) Licensee's actions during outage to clean boric acid deposits from the vessel bottom head to establish a baseline for future inspections	Vessel surface and the annulus regions of the RPV lower head penetrations were power washed to establish a baseline for future inspections.

Table 5.3-2, Inspection Results, Fall 2003 Outage

Inspection Attributes\Plant	Byron Station, Unit 1
(1) Type and extent of inspection	All 58 BMI penetrations were examined by VT-2 qualified inspectors, 360^0 around each circumference with a video camera.
(2) Identification of boric acid deposits and characterization of deposits	The licensee identified boric acid trails with no discernable thickness from borated water looking down the side of the RPV. The licensee postulated that these trails originated from the previous reactor cavity seal leakage. The licensee also identified a minor flaky corrosion surface layer with no discernable thickness on the RPV lower head.
(3) Chemical analysis, number of samples of the deposit, and results	The licensee did not perform chemical analysis.
(4) Licensee's actions during outage to clean boric acid deposits from the vessel bottom head to establish a baseline for future inspections	The licensee cleaned the lower head using a low pressure power wash to establish a baseline for future inspections.

Table 5.3-2, Inspection Results, Fall 2003 Outage

Inspection Attributes\Plant	Catawba, Unit 1
(1) Type and extent of inspection	Using digital cameras and direct visual observation, a 360^0 inspection was performed on 100 percent of the BMI penetrations.
(2) Identification of boric acid deposits and characterization of deposits	The licensee did not observe any white boric acid deposits that were indicative of BMI leakage. The licensee identified rust trails and scaling on the bottom of the RPV. Based on the appearance of the rust trails, the licensee concluded that the source for these rust trails was refuel cavity seal leakage.
(3) Chemical analysis, number of samples of the deposit, and results	The licensee obtained an isotopic analysis of the rust trails. The licensee did not analyze the samples for the presence of boron and lithium.
(4) Licensee's actions during outage to clean boric acid deposits from the vessel bottom head to establish a baseline for future inspections	The bare metal surface of the RPV was cleaned and re-inspected to establish a baseline for future inspections.

Table 5.3-2, Inspection Results, Fall 2003 Outage

Inspection Attributes\Plant	Comanche Peak, Unit 2
(1) Type and extent of inspection	All 58 BMI penetrations were examined by VT-2 qualified inspectors, 360^0 around each circumference with a video camera attached to a robotic crawler.
(2) Identification of boric acid deposits and characterization of deposits	The licensee identified white markings with no discernable thickness at some of the RPV lower head penetrations. Based on the appearance and texture of these markings, the licensee stated that there was no evidence of boric acid leakage at the RPV lower head penetrations.
(3) Chemical analysis, number of samples of the deposit, and results	The licensee did not perform chemical analysis.
(4) Licensee's actions during outage to clean boric acid deposits from the vessel bottom head to establish a baseline for future inspections	Not addressed.

Table 5.3-2, Inspection Results, Fall 2003 Outage

Inspection Attributes\Plant	D. C. Cook, Unit 1
(1) Type and extent of inspection	All 58 BMI penetrations on the RPV lower head were examined 360^0 around each circumference with a video camera attached to an inspection pole. Visual examination of the RPV lower head area was performed by VT-2 inspectors.
(2) Identification of boric acid deposits and characterization of deposits	The licensee's initial inspection results revealed an area of apparent boric acid flow originating above the insulation support ring and flowing toward the center of the vessel bottom. This area was located in a single quadrant and had a streaked appearance consisting of rust staining and dry crystalline deposits, presumed to be dry boric acid.
(3) Chemical analysis, number of samples of the deposit, and results	Chemical analysis was performed on the observed boric acid residue that was present in the rust trails on the RPV head. The licensee concluded from the results that the source of the material was leakage from the reactor refueling cavity. The licensee observed no boric acid deposits of discernable thickness on the vessel lower head. In addition, the rust trails appeared to have originated from above the vessel lower head. Thus, the licensee concluded that there was no RCS leakage at the RPV lower head penetrations.
(4) Licensee's actions during outage to clean boric acid deposits from the vessel bottom head to establish a baseline for future inspections	The licensee cleaned the RPV lower head using warm water washing.

Table 5.3-2, Inspection Results, Fall 2003 Outage

Inspection Attributes\Plant	Crystal River, Unit 3
(1) Type and extent of inspection	All 52 BMI penetrations were examined 360^0 around each circumference by VT-2 qualified inspectors. The results of the examination have been recorded on visual examination sheets and sent to records for retention.
(2) Identification of boric acid deposits and characterization of deposits	The licensee identified loose rusty scale and flaking paint on the RPV lower head. The licensee did not observe any white boric acid residue on the RPV lower head.
(3) Chemical analysis, number of samples of the deposit and results	The licensee did not perform chemical analysis.
(4) Licensee's actions during outage to clean boric acid deposits from the vessel bottom head to establish a baseline for future inspections	Not addressed.

Table 5.3-2, Inspection Results, Fall 2003 Outage

Inspection Attributes\Plant	Ginna
(1) Type and extent of inspection	All 36 BMI penetrations were examined 360° around each circumference with a video camera having VT-1 quality resolution. The camera was attached to an adjustable inspection pole. Each penetration had unique identification marking which enabled the confirmation of 100% inspection. All Inspections were performed by VT-2 inspectors and verified by a Level III inspector.
(2) Identification of boric acid deposits and characterization of deposits	Examination indicated broad diffused boric acid residue having the appearance of an opaque film in several areas around the bottom head. The residue had no buildup or deposit thickness.
(3) Chemical analysis, number of samples of the deposit and results	Samples for isotopic analysis of the residue were taken from three penetrations, and two areas of base metal to establish the age of the boric acid residue. There were a number of possible sources for these radioactive species. The licensee did not or was not able to collect and analyze samples for the presence of boron and lithium. However, the licensee observed no boric acid deposits of discernable thickness on the bottom head. On this basis of observation, the licensee concluded that there was no leakage at the RPV lower head penetrations.
(4) Licensees actions during outage to clean boric acid deposits from the vessel bottom head to establish a baseline for future inspections	The licensee had steam cleaned 3" around each penetration with de-ionized water and stainless steel wire brushes to establish a baseline for future inspections.

Table 5.3-2, Inspection Results, Fall 2003 Outage

Inspection Attributes\Plant	McGuire, Unit 2
(1) Type and extent of inspection	All BMI penetrations were examined 360^0 around each circumference with a video camera, and direct visual observations.
(2) Identification of boric acid deposits and characterization of deposits	The licensee identified dried borated water trails running down the side and across the bottom of the reactor vessel. Trails intersected the BMI penetrations creating thin boron acid deposits. The licensee stated that none of the deposits were characteristic of through wall leakage.
(3) Chemical analysis, number of samples of the deposit and results	Isotopic analysis of samples taken at annulus of randomly selected penetrations indicated the age of deposits to be greater than 18 months. The licensee did not or was not able to collect and analyze samples for the presence of boron and lithium. Therefore, the chemical analysis was of limited value. However, the licensee observed no boric acid deposits of discernable thickness on the lower head. On this basis of observation, the licensee concluded that there was no leakage at the RPV lower head penetrations.
(4) Licensee's actions during outage to clean boric acid deposits from the vessel lower head to establish a baseline for future inspections	Bare metal surface of the RPV lower head was cleaned and re-inspected prior to restart to establish baseline for future inspection.

Table 5.3-2, Inspection Results, Fall 2003 Outage

Inspection Attributes\Plant	Oconee, Unit 1
(1) Type and extent of inspection	The licensee conducted BMV inspection of the RPV lower head, including 360^0 around 100% of the bottom BMI penetrations. The licensee performed this inspection using video cameras and direct visual observation.
(2) Identification of boric acid deposits and characterization of deposits	The inspection showed no evidence of boron or other indications of leakage from the RPV lower head or penetrations. However, the licensee observed a loose and flaky protective coating on the RPV lower head.
(3) Chemical analysis, number of samples of the deposit, and results	The licensee did not perform chemical analysis.
(4) Licensee's actions during outage to clean boric acid deposits from the vessel lower head to establish a baseline for future inspections	The licensee cleaned the lower head to establish a baseline for future inspections.

Table 5.3-2, Inspection Results, Fall 2003 Outage

Inspection Attributes\Plant	Palo Verde, Unit 2
(1) Type and extent of inspection	All 61 BMI penetrations were examined 360^0 around each circumference with a video camera having zoom capabilities attached to a robot. Examinations were performed by Level III, VT-2 qualified inspectors.
(2) Identification of boric acid deposits and characterization of deposits	The licensee did not observe boric acid deposits in the annulus region. The licensee identified streaks and stains on the outside of the RPV lower head. The licensee concluded that the staining was caused by the spillage from the control rod drive mechanism air conditioning units that leaked during previous refueling outage. The licensee also observed spraylat (trade name) which is a protective coating that was used during construction and tape materials on the RPV lower head.
(3) Chemical analysis, number of samples of the deposit, and results	The licensee did not perform chemical analysis.
(4) Licensee's actions during outage to clean boric acid deposits from the vessel lower head to establish a baseline for future inspections	The licensee cleaned the spraylat coating and tape from the annulus area of the penetrations using dry ice media. The cleaning process provided a clean zone of ½" around the penetration annulus areas. A bare metal zone was achieved on 39 of 61 penetrations before equipment problems occurred. The licensee cleaned the remaining 22 penetrations during the spring 2005 refueling outage.

Table 5.3-2, Inspection Results, Fall 2003 Outage

43

Inspection Attributes\Plant	Point Beach, Unit 2
(1) Type and extent of inspection	The licensee examined each of the 36 RPV BMI penetrations 360^0 around the circumference using digital cameras. In addition, the RPV lower head surface was inspected up to 6 inches above the highest BMI penetration. The visual examination was accomplished utilizing VT-2 certified personnel.
(2) Identification of boric acid deposits and characterization of deposits	Minor rust staining with no discernable thickness was observed on the side of the RPV. The licensee determined that these trails originated from previous reactor cavity seal leakage, because they had no volume. The licensee found white residue on the penetration to tube weld region of multiple BMI penetrations. The licensee evaluated this residue and determined it to be liquid dye penetrant developer. The licensee stated that this developer had been left on the RPV since original construction.
(3) Chemical analysis, number of samples of the deposit, and results	The licensee did not perform chemical analysis.
(4) Licensee's actions during outage to clean boric acid deposits from the vessel lower head to establish a baseline for future inspections	Not addressed.

Table 5.3-2, Inspection Results, Fall 2003 Outage

Inspection Attributes\Plant	Prairie Island, Unit 2
(1) Type and extent of inspection	12 penetrations were inspected using a hand held video probe, the remaining 24 were inspected using the video probe mounted on the vessel. VT-2 inspectors were used. The licensee believes that all the circumferences of 36 penetrations were examined, although due to minor access limitations some small sections of the annulus may not have been directly observed.
(2) Identification of boric acid deposits and characterization of deposits	The licensee observed liquid streaks with white powdery consistency that emanated from above the penetrations and ran down to the bottom of the RPV and impinged on four penetrations.
(3) Chemical analysis, number of samples of the deposit and results	Samples of the deposit indicated that boron and lithium were at or below the lower delectability limits. The licensee observed no boric acid deposits of discernable thickness on the RPV lower head. On this basis of these results and observations, the licensee concluded that there were no leaking lower head penetrations.
(4) Licensee's actions during outage to clean boric acid deposits from the vessel lower head to establish a baseline for future inspections	Not addressed.

Table 5.3-2, Inspection Results, Fall 2003 Outage

Inspection Attributes\Plant	Salem, Unit 2
(1) Type and extent of inspection	All 58 BMI penetrations on the RPV lower head were examined with a video camera. The visual examination was performed by VT-2 inspectors and Level III examiners.
(2) Identification of boric acid deposits and characterization of deposits	The licensee observed small amounts of white crystallized substances on the insulation below the RPV lower head. In addition, translucent and rust trails were found on the RPV lower head. The licensee concluded that these trails had originated from either the previous reactor cavity seal leak, or RPV head canopy seal leak cleanup in 1987, or RCS system hot leg sample valve cleanup.
(3) Chemical analysis, number of samples of the deposit and results	Chemical analysis was performed of the observed boric acid residue that is present in the rust trails on the RPV lower head. The results confirmed that the source of the material was leakage that occurred during operations from the RCS system hot leg sample valve clean up. The licensee observed no boric acid deposits of discernable thickness on the RPV lower head. In addition, the rust trails appeared to have originated from above the vessel lower head. On this basis of observation, the licensee concluded that there was no leakage at the RPV lower head penetrations.
(4) Licensee's actions during outage to clean boric acid deposits from the vessel lower head to establish a baseline for future inspections	Not addressed.

Table 5.3-2, Inspection Results, Fall 2003 Outage

Inspection Attributes\Plant	Surry, Unit 2
(1) Type and extent of inspection	The licensee performed 360^0 visual examination of the 50 RPV lower head penetrations. This inspection was conducted using either direct visual inspection or visual inspection aided by mirrors. The licensee took digital photographs to record portions of the inspection.
(2) Identification of boric acid deposits and characterization of deposits	The licensee stated that it did not find evidence of leakage or wastage.
(3) Chemical analysis, number of samples of the deposit, and results	The licensee did not perform chemical analysis.
(4) Licensee's actions during outage to clean boric acid deposits from the vessel lower head to establish a baseline for future inspections	Even though there was no evidence of wastage or boric acid residue on the RPV lower head, the licensee cleaned the RPV lower head to establish a baseline for future inspections.

Table 5.3-2, Inspection Results, Fall 2003 Outage

47

Inspection Attributes\Plant	Seabrook
(1) Type and extent of inspection	All 58 BMI penetrations on the RPV lower head were examined 360^0 around each circumference with a video camera. Direct visual inspection was supplemented with a mirror, drop lights and flashlights. Certified VT-2 visual examiners conducted the inspection.
(2) Identification of boric acid deposits and characterization of deposits	The licensee observed a few boric acid streams originating from above the lower head and some tape residue on the lower head. The licensee assumed that the tape was applied to penetrations as a protective cover during construction.
(3) Chemical analysis, analysis number of samples of the deposit and results	The licensee performed a chemical analysis of the boric acid residue observed on the RPV lower head. Based on the chemical analysis, the licensee noted that the boric acid residue may have originated during the prior cavity seal ring leakage which occurred before 1996. The licensee observed no boric acid deposits of discernable thickness on the vessel lower head. In addition, the boric acid appeared to have originated from above the vessel lower head. On the basis of these observations, the licensee concluded that there was no leakage at the RPV lower head penetrations.
(4) Licensee's actions during outage to clean boric acid deposits from the vessel lower head to establish a baseline for future inspections	The licensee used demineralized water to clean the boric acid streams.

Table 5.3-2, Inspection Results, Fall 2003 Outage

Inspection Attributes\Plant	Sequoyah, Unit 2
(1) Type and extent of inspection	All 58 BMI penetrations on the RPV lower head were examined 360^0 around each circumference with a video camera using video equipment mounted on remote magnetic crawlers. Each penetration was viewed in 2 separate 180^0 segments to ensure 100% coverage. The visual examination was performed by Level III examiners.
(2) Identification of boric acid deposits and characterization of deposits	The licensee observed boric acid deposits that had no discernable thickness on the RPV lower head. Due to the tightly adhering nature of the deposits, chemical samples were not taken from these deposits. However, swipe tests were performed on these deposits. On 25 BMI penetrations, the licensee found a dark tar-like substance that originated about 1 to 2 inches below where the penetrations exit the vessel.
(3) Chemical analysis, number of samples of the deposit and results	The thin boric acid appeared to have originated from above the vessel lower head. Based on the swipe test results of the boric acid deposits, the licensee noted that the boric acid residue may have originated during the prior cavity seal ring leakage and not from the lower head penetrations. The licensee indicated that there was no boron leakage from any of the 58 BMI penetrations. The licensee believed that the likely source of the dark tar-like substance was the flagging tape that was used for identification purposes during the original construction. Chemical analysis of the dark residue of the tape showed presence of some fluorides, chlorides, and sulfates. In order to establish that stress corrosion cracking was not an issue, liquid penetrant tests were performed on three BMI penetrations and there were no recordable indications.
(4) Licensee's actions during outage to clean boric acid deposits from the vessel lower head to establish a baseline for future inspections	Not addressed.

Table 5.3-2, Inspection Results, Fall 2003 Outage

Inspection Attributes\Plant	V. C. Summer
(1) Type and extent of inspection	The licensee performed a BMV inspection of the lower RPV head surface and RPV lower head penetrations. The inspection covered 360^0 around each penetration on the RPV lower head. Part of this inspection was done with a camera, while the more accessible areas were inspected visually. The remote inspection was performed using a robotic camera. The record produced from the robotic camera was reviewed and evaluated by qualified VT-2, Level II personnel.
(2) Identification of boric acid deposits and characterization of deposits	The licensee found some dried boric acid residue on the bottom and sides of the RPV. This boric acid originated from above and ran down the sides of the RPV onto the bottom RPV lower head. There were about twenty penetrations that had white residue bridging the narrow gap next to the penetrations. Additional inspections determined that there was no buildup of boric acid in the annulus region. The licensee found boric acid residue on the reactor vessel insulation, the in-core pit areas, and on the floor below the RPV. The licensee attempted to determine the source of the dried boric acid residue using remote visual aids; however, obstructions in the annulus between the vessel wall and the insulation prevented a complete inspection up the side of the vessel.
(3) Chemical analysis, analysis number of samples of the deposit and results	The licensee performed a chemical analysis of the boric acid residue observed on the RPV lower head. Based on the analysis, the licensee noted that the boric acid residue may have originated during the prior cavity seal ring leakage. The licensee also indicated that the presence of boric acid may have been due to accumulations associated with "A" hot leg through wall flaw that was discovered in 2001. The licensee observed no boric acid deposits of discernable thickness on the vessel lower head. In addition, the boric acid appeared to have originated from above the vessel lower head. On the basis of these observations, the licensee concluded that there was no leakage at the lower head penetrations.
(4) Licensee's actions during outage to clean boric acid deposits from the vessel lower head to establish a baseline for future inspections	The licensee removed some of the boric acid deposits from the RPV lower head to establish a baseline for future inspections.

Table 5.3-2, Inspection Results, Fall 2003 Outage

Inspection Attributes\Plant	Three Mile Island, Unit 1
(1) Type and extent of inspection	All 52 BMI penetrations were examined by VT-2 qualified inspectors, 360^0 around each circumference with a remote video camera attached to a robotic crawler.
(2) Identification of boric acid deposits and characterization of deposits	The licensee observed traces of boric acid on the RPV surface adjacent to an insulation skirt access hole that allowed leakage from the fuel transfer canal seal plate during previous refueling outage. These boron traces were semi-transparent deposits with no discernable thickness. The licensee indicated that there was no RCS leakage in the BMI penetrations. Furthermore, the licensee did not find any base metal wastage in the RPV lower head.
(3) Chemical analysis, number of samples of the deposit, and results	The licensee did not perform chemical analysis.
(4) Licensee's actions during outage to clean boric acid deposits from the vessel lower head to establish a baseline for future inspections	The licensee cleaned the RPV lower head using wet cloths and scotch brite pads to establish a baseline for future inspections.

Table 5.3-2, Inspection Results, Fall 2003 Outage

Inspection Attributes\Plant	Turkey Point, Unit 4
(1) Type and extent of inspection	All 50 BMI penetrations on the RPV lower head were examined 360^0 around each circumference with a remote camera mounted on magnetic crawler or a camera mounted on a long handle pole. The camera mounted on a long handle pole was used to perform inspections on penetrations with limited access. The visual examination was performed by Level II inspectors.
(2) Identification of boric acid deposits and characterization of deposits	The licensee observed a thin dry translucent film around some BMI penetrations and RPV lower head. The licensee postulated that the possible source of the film was prior cavity seal ring leakage or wash down events.
(3) Chemical analysis, number of samples of the deposit and results	The licensee performed a chemical analysis on two representative BMI penetrations. No lithium or boron was detected in these samples, which would indicate that the leakage was not from the RCS. Other chemical analyses performed were of limited value. The licensee observed no boric acid deposits of discernable thickness on the RPV lower head. Based on these results, the licensee concluded that the presence of the residue was not due to RCS leakage.
(4) Licensee's actions during outage to clean boric acid deposits from the vessel lower head to establish a baseline for future inspections	Not addressed.

Table 5.3-2, Inspection Results, Fall 2003 Outage

Inspection Attributes\Plant	Vogtle, Unit 1
(1) Type and extent of inspection	All 58 BMI penetrations on the RPV lower head were examined 360^0 around each circumference.
(2) Identification of boric acid deposits and characterization of deposits	The licensee found a V shaped rust stain that was oriented below loop 3 cold leg and extended from bottom of the RPV up the side about 10 feet.
(3) Chemical analysis, analysis number of samples of the deposit and results	The licensee performed a chemical analysis on the rust stains around BMI and loop 3 cold leg areas. The analysis indicated low levels of boron, and below detectable limits of lithium. Based on the analysis, the licensee postulated that the boric acid residue may have originated during the prior cavity seal ring leakage. The licensee observed no boric acid deposits of discernable thickness on the vessel lower head. Based on these observations, the licensee concluded that there was no leakage at the RPV lower head penetrations.
(4) Licensee's actions during outage to clean boric acid deposits from the vessel lower head to establish a baseline for future inspections	Not addressed.

Table 5.3-2, Inspection Results, Fall 2003 Outage

Inspection Attributes\Plant	Watts Bar, Unit 1
(1) Type and extent of inspection	All 58 BMI penetrations were examined by Level III certified inspectors, 360^0 around each circumference with a high resolution video camera attached to a robotic crawler.
(2) Identification of boric acid deposits and characterization of deposits	The licensee identified a minor flaky corrosion surface layer with no discernable thickness at the RPV lower head. Light to moderate staining and surface rust was seen around the annular area and on the bare metal surface. These areas had trails from above the lower head. The licensee determined that these trails were not associated with boric acid leakage from the BMI penetrations.
(3) Chemical analysis, number of samples of the deposit, and results	The licensee did not perform chemical analysis.
(4) Licensee's actions during outage to clean boric acid deposits from the vessel lower head to establish a baseline for future inspections	Not addressed.

Table 5.3-2, Inspection Results, Fall 2003 Outage

Inspection Attributes\Plant	Wolf Creek
(1) Type and extent of inspection	All BMI penetrations were examined by VT-2 qualified Level II inspectors, 360° around each circumference with a video camera.
(2) Identification of boric acid deposits and characterization of deposits	The licensee identified trails of boric acid staining on the side of the RPV. The licensee postulated that these trails originated from previous reactor cavity seal leakage. The licensee also identified a minor flaky corrosion surface layer with no discernable thickness on the RPV lower head. The licensee found no material wastage and did not identify any RCS leakage through the BMI penetrations.
(3) Chemical analysis, number of samples of the deposit, and results	The licensee did not perform chemical analysis.
(4) Licensee's actions during outage to clean boric acid deposits from the vessel lower head to establish a baseline for future inspections	The licensee cleaned the lower head and residues around three penetrations to establish a baseline for future inspections.

Table 5.3-2, Inspection Results, Fall 2003 Outage

Inspection Attributes\Plant	Arkansas Nuclear, Unit 1
(1) Type and extent of inspection	The licensee visually inspected all 52 BMI penetrations including 100% of the circumference of each penetration annulus region.
(2) Identification of boric acid deposits and characterization of deposits	The licensee did not identify any boric acid deposits on the RPV lower head surface. Based on the inspection results, the licensee concluded that there was no RCS leakage from the BMI penetrations.
(3) Chemical analysis, number of samples of the deposit, and results	The licensee did not perform chemical analysis.
(4) Licensee's actions during outage to clean boric acid deposits from the vessel lower head to establish a baseline for future inspections	Not addressed.

Table 5.3-3, Inspection Results, Spring 2004 Outage

Inspection Attributes\Plant	Byron Station, Unit 2
(1) Type and extent of inspection	All 58 BMI penetrations on the RPV lower head were examined with a remote camera. The visual examination was performed by VT-2 qualified inspectors.
(2) Identification of boric acid deposits and characterization of deposits	The licensee did not identify any boric acid deposits in the annulus region between the penetration and the reactor vessel. The licensee observed several rust trails on the RPV lower head surface. The licensee indicated that these trails originated from the previous reactor cavity seal.
(3) Chemical analysis, number of samples of the deposit, and results	The licensee did not perform chemical analysis.
(4) Licensee's actions during outage to clean boric acid deposits from the vessel lower head to establish a baseline for future inspections	Not addressed.

Table 5.3-3, Inspection Results, Spring 2004 Outage

Inspection Attributes\Plant	Callaway, Unit 1
(1) Type and extent of inspection	The licensee visual inspected 58 BMI penetrations including 100% of the circumference of each penetration annulus region. In addition, the licensee performed ultrasonic and eddy current examinations on these BMI penetrations.
(2) Identification of boric acid deposits and characterization of deposits	The licensee identified thin white stains on the reactor vessel lower head. Based on its appearance, the licensee concluded that these stains originated from above the lower head. The licensee observed debris on the annulus regions of the BMI penetrations 4 and 6. The licensee stated that the debris had the appearance of fine brown color. The licensee cleaned the debris in the annulus regions of the BMI penetrations 4 and 6, and performed ultrasonic and eddy current examinations of these areas, and found no indications that could have resulted in leakage. The licensee stated that the results of the ultrasonic and eddy current examinations revealed no cracking or significant lack of fusion in any of the 58 penetrations and their associated J-groove welds. Based on these observations, the licensee concluded that there was no RCS leakage on the RPV lower head.
(3) Chemical analysis, number of samples of the deposit, and results	The licensee did not perform chemical analysis.
(4) Licensee's actions during outage to clean boric acid deposits from the vessel lower head to establish a baseline for future inspections	The licensee cleaned the debris in the annulus regions of the BMI penetrations 4 and 6.

Table 5.3-3, Inspection Results, Spring 2004 Outage

Inspection Attributes\Plant	Comanche Peak, Unit 1
(1) Type and extent of inspection	Visual inspection of 50 BMI penetrations including 100% of the circumference of each penetration annulus was done using a remote crawler with a video camera. The crawler traversed on the horizontal surface of the insulation panel. The visual examination was documented on a video tape.
(2) Identification of boric acid deposits and characterization of deposits	The licensee did not identify any boric acid residue on the reactor vessel lower head. However, the licensee observed a few rust trails originating from above the lower head. The licensee stated that these rust trails did not have any discernable thickness. Based on the appearance of the rust trails, the licensee concluded that there was no RCS leakage on the reactor vessel lower head.
(3) Chemical analysis, number of samples of the deposit, and results	The licensee did not perform chemical analysis.
(4) Licensee's actions during outage to clean boric acid deposits from the vessel lower head to establish a baseline for future inspections	Not addressed.

Table 5.3-3, Inspection Results, Spring 2004 Outage

Inspection Attributes\Plant	Diablo Canyon, Unit 1
(1) Type and extent of inspection	Visual inspection of 58 BMI penetrations including 100% of the circumference of each penetration annulus was done using a remote crawler with a video camera. The visual examination was performed by VT-2 qualified inspectors, and was documented on a video tape.
(2) Identification of boric acid deposits and characterization of deposits	The licensee did not identify any boric acid accumulation on the BMI penetrations or on the RPV lower head. However, the licensee observed slight trails which appear to be boric acid residue. The licensee stated that these trails did not have any discernable thickness. The licensee observed similar trails on the outside vessel insulation and on the concrete bio-shield wall. The licensee postulated that these trails originated from the previous reactor cavity seal. Based on the appearance of these trails, the licensee concluded that there was no RCS leakage on the reactor vessel lower head.
(3) Chemical analysis, number of samples of the deposit, and results	The licensee did not perform chemical analysis.
(4) Licensee's actions during outage to clean boric acid deposits from the vessel lower head to establish a baseline for future inspections	Not addressed.

Table 5.3-3, Inspection Results, Spring 2004 Outage

Inspection Attributes\Plant	Joseph M. Farley, Unit 2
(1) Type and extent of inspection	The licensee visually inspected all 50 BMI penetrations.
(2) Identification of boric acid deposits and characterization of deposits	The licensee did not identify any boric acid residue on the reactor vessel lower head. However, the licensee observed a few light rust stains and tape residue on the reactor vessel lower head. The licensee indicated that it did not identify any evidence of RPV lower head material wastage or any RCS leakage.
(3) Chemical analysis, number of samples of the deposit, and results	The licensee did not perform chemical analysis.
(4) Licensee's actions during outage to clean boric acid deposits from the vessel lower head to establish a baseline for future inspections	Based on the visual examination of the reactor vessel lower head, the licensee determined not to clean the reactor vessel lower head.

Table 5.3-3, Inspection Results, Spring 2004 Outage

Inspection Attributes\Plant	McGuire, Unit 1
(1) Type and extent of inspection	All BMI penetrations were examined 360^0 around each circumference with a video camera, and direct visual observation.
(2) Identification of boric acid deposits and characterization of deposits	The licensee identified some translucent boron and rust-like deposits on the reactor vessel lower head. Based on the appearance and thickness of the deposits, the licensee concluded that the source of these deposits was not due RCS leakage. The licensee performed isotopic analysis to confirm that these deposits did not originate from the RCS leakage.
(3) Chemical analysis, number of samples of the deposit, and results	The licensee performed isotopic analysis of the deposits and confirmed that the origin of the deposits was the reactor cavity seal leakage from prior refueling outage.
(4) Licensee's actions during outage to clean boric acid deposits from the vessel lower head to establish a baseline for future inspections	The licensee cleaned the reactor vessel lower head to establish a baseline for future inspections.

Table 5.3-3, Inspection Results, Spring 2004 Outage

Inspection Attributes\Plant	Millstone, Unit 3
(1) Type and extent of inspection	All 58 BMI penetrations on the RPV lower head were examined 360^0 around each circumference by VT-2 qualified inspectors.
(2) Identification of boric acid deposits and characterization of deposits	The licensee did not observe any residue in the annulus region of the penetration and the reactor vessel lower head. The licensee identified a thin transparent film on the RPV lower head. The licensee indicated that this film originated due to reactor cavity seal leakage from an area above the lower head penetrations. The licensee also indicated that it did not identify any evidence of RPV lower head material wastage.
(3) Chemical analysis, number of samples of the deposit, and results	The licensee did not perform chemical analysis.
(4) Licensee's actions during outage to clean boric acid deposits from the vessel lower head to establish a baseline for future inspections	The RPV lower head was cleaned to establish a baseline for future inspections.

Table 5.3-3, Inspection Results, Spring 2004 Outage

Inspection Attributes\Plant	North Anna, Unit 2
(1) Type and extent of inspection	All 50 BMI penetrations were examined by VT-2 qualified inspectors, 360^0 around each circumference with a video camera.
(2) Identification of boric acid deposits and characterization of deposits	The licensee identified some minor rust stains and peeling of a paint on the reactor vessel lower head. The licensee did not identify any boric acid deposits.
(3) Chemical analysis, number of samples of the deposit, and results	The licensee did not perform chemical analysis.
(4) Licensee's actions during outage to clean boric acid deposits from the vessel lower head to establish a baseline for future inspections	Not addressed.

Table 5.3-3, Inspection Results, Spring 2004 Outage

Inspection Attributes\Plant	Oconee, Unit 2
(1) Type and extent of inspection	All BMI penetrations were examined 360^0 around each circumference with a video camera, and direct visual observation.
(2) Identification of boric acid deposits and characterization of deposits	The licensee identified some minor flaking of the original coating on the reactor vessel lower head. The licensee did not identify any boric acid deposits or any wastage on the reactor vessel lower head.
(3) Chemical analysis, number of samples of the deposit, and results	The licensee did not perform chemical analysis.
(4) Licensee's actions during outage to clean boric acid deposits from the vessel lower head to establish a baseline for future inspections	The licensee cleaned the degraded coating on the reactor vessel lower head to establish a baseline for future inspections.

Table 5.3-3, Inspection Results, Spring 2004 Outage

Inspection Attributes\Plant	Palo Verde, Unit 1
(1) Type and extent of inspection	All 61 BMI penetrations were examined 360^0 around each circumference with a video camera having zoom capabilities attached to a robot. These inspections were conducted by level III, VT-2 qualified inspectors.
(2) Identification of boric acid deposits and characterization of deposits	The licensee did not observe boric acid deposits in the annulus region. However, small flakes of spraylat (trade name) coating was observed in the annulus region of some of the BMI penetrations. The licensee also identified small amounts of dry red oxide deposits at the annulus regions on one-third of BMI penetrations. The licensee concluded that the source of these deposits was the leakage from the control rod drive mechanism air conditioning units that occurred during the previous refueling outage.
(3) Chemical analysis, number of samples of the deposit, and results	The licensee did not perform chemical analysis.
(4) Licensee's actions during outage to clean boric acid deposits from the vessel lower head to establish a baseline for future inspections	The licensee determined that cleaning of the reactor vessel lower head was not necessary.

Table 5.3-3, Inspection Results, Spring 2004 Outage

Inspection Attributes\Plant	Point Beach, Unit 1
(1) Type and extent of inspection	All 36 BMI penetrations were examined by VT-2 qualified Inspectors around each circumference with a video camera.
(2) Identification of boric acid deposits and characterization of deposits	The licensee did not observe boric acid deposits in the annulus region. However, the licensee observed several rust trails with no discernable thickness on the RPV lower head surface. The licensee postulated that these trails originated from the previous reactor cavity seal.
(3) Chemical analysis, number of samples of the deposit, and results	The licensee did not perform chemical analysis.
(4) Licensee's actions during outage to clean boric acid deposits from the vessel lower head to establish a baseline for future inspections	Not addressed.

Table 5.3-3, Inspection Results Spring 2004 Outage

Inspection Attributes\Plant	H. B. Robinson, Unit 2
(1) Type and extent of inspection	All 50 BMI penetrations were examined with a remote camera by level II and Level III VT-2 qualified inspectors.
(2) Identification of boric acid deposits and characterization of deposits	The licensee identified rust and boric acid residue on the reactor vessel lower head. This was attributed to the previous reactor cavity seal leakage. The licensee indicated that it did not identify any evidence of RPV lower head material wastage or any RCS leakage.
(3) Chemical analysis, number of samples of the deposit, and results	The licensee did not perform chemical analysis.
(4) Licensee's actions during outage to clean boric acid deposits from the vessel lower head to establish a baseline for future inspections	The licensee cleaned the lower head to establish a baseline for future inspections.

Table 5.3-3, Inspection Results, Spring 2004 Outage

Inspection Attributes\Plant	Salem, Unit 1
(1) Type and extent of inspection	All 58 BMI penetrations were examined by level III, VT-2 qualified inspectors, 360^0 around each circumference of the penetration.
(2) Identification of boric acid deposits and characterization of deposits	The licensee did not observe boric acid deposits in the annulus region. However, the licensee observed rust trails on 40 of the 58 BMI penetrations. The licensee identified a white translucent residue with no discernable thickness on five of these 40 BMI penetrations. The licensee stated that one of the 40 BMI penetrations had a white translucent residue at the annulus region. Based on these observations, the licensee concluded that the residue originated from the previous reactor cavity seal leakage.
(3) Chemical analysis, number of samples of the deposit, and results	The licensee did not perform chemical analysis.
(4) Licensee's actions during outage to clean boric acid deposits from the vessel lower head to establish a baseline for future inspections	The licensee washed the lower head to establish a baseline for future inspections.

Table 5.3-3, Inspection Results, Spring 2004 Outage

Inspection Attributes\Plant	South Texas, Unit 2
(1) Type and extent of inspection	All 58 BMI penetrations on the RPV lower head were examined with a remote camera. The visual examination was performed by VT-2 inspectors.
(2) Identification of boric acid deposits and characterization of deposits	The licensee identified deposits in the annulus region of seven penetrations. Based on the visual appearance, the licensee postulated that these deposits could be either a coating or a sealant or adhesive residue. The licensee stated that the surface of the RPV lower head appeared to be clean.
(3) Chemical analysis, number of samples of the deposit, and results	Licensee performed a chemical analysis on four representative BMI penetrations. Low concentrations of lithium (<0.005%), and boron(<0.15%) were detected. Traces of Cobalt 60 was found in the deposits. Cesium 137 was observed in one of the samples. The licensee stated that these radioactive species can originate from a number of sources other than RCS leakage from the vessel. Based on the low concentration levels of boron and lithium, the licensee concluded that the source of the observed deposits is not from RCS leakage.
(4) Licensee's actions during outage to clean boric acid deposits from the vessel lower head to establish a baseline for future inspections	The licensee cleaned the seven penetrations to establish a baseline for future inspections.

Table 5.3-3, Inspection Results, Spring 2004 Outage

Inspection Attributes\Plant	Vogtle, Unit 2
(1) Type and extent of inspection	Each circumference of all BMI penetrations was examined. The licensee did not provide details on the qualification of inspectors and the method of bare metal visual examination of the RPV lower head.
(2) Identification of boric acid deposits and characterization of deposits	The licensee did not identify any boric acid deposits or any material wastage on the RPV lower head.
(3) Chemical analysis, number of samples of the deposit, and results	The licensee did not perform chemical analysis.
(4) Licensee's actions during outage to clean boric acid deposits from the vessel lower head to establish a baseline for future inspections	Not addressed.

Table 5.3-3, Inspection Results, Spring 2004 Outage

Inspection Attributes\Plant	Beaver Valley, Unit 1
(1) Type and extent of inspection	The licensee visually inspected all 50 BMI penetrations including 100% of each penetration using a camera to record the inspection. The visual examination was performed by VT-2 Level II qualified inspectors and approved by Level III inspectors.
(2) Identification of boric acid deposits and characterization of deposits	The licensee identified boric acid stains with no appreciable volume on the RPV lower head surface. The licensee indicated that the trails originated from the previous reactor cavity seal leakage. The licensee did not identify any through wall leakage from any BMI penetrations.
(3) Chemical analysis, number of samples of the deposit, and results	The licensee did not perform chemical analysis.
(4) Licensee's actions during outage to clean boric acid deposits from the vessel lower head to establish a baseline for future inspections	Even though the licensee did not observe any boric acid deposits on the lower head, it cleaned and inspected the lower head to establish a baseline for future inspections.

Table 5.3-4, Inspection Results, Fall 2004 Outage

Inspection Attributes\Plant	Braidwood Station, Unit 1
(1) Type and extent of inspection	The licensee visually inspected all 58 BMI penetrations including 360^0 of each penetration using a remote camera to record the inspection. The visual examination was performed by VT-2 inspectors Level II qualified inspectors.
(2) Identification of boric acid deposits and characterization of deposits	The licensee did not identify any boric acid deposits however, it observed minor accumulation of debris at penetrations 38, 43, 44, 45 and 52 with thickest deposit at the penetration 44. In addition, the licensee observed minor corrosion of RPV lower head surface.
(3) Chemical analysis, number of samples of the deposit, and results	In order to establish the source of the debris at the five penetrations, the licensee performed chemical analysis and Gamma Spectroscopy and concluded that the material in the debris came from the insulation. The radionuclides that were present in the samples indicated that there was no RCS leakage. Visual examination of the five penetrations before and after taking the samples indicated that the debris was not located at the annulus region around the penetrations. Based on these observations and analyses, the licensee concluded that there was no RCS leakage at these penetrations.
(4) Licensee's actions during outage to clean boric acid deposits from the vessel lower head to establish a baseline for future inspections	Not addressed.

Table 5.3-4, Inspection Results, Fall 2004 Outage

Inspection Attributes\Plant	Catawba, Unit 2
(1) Type and extent of inspection	The licensee visually inspected all 58 BMI penetrations including 360^0 of each penetration using a remote camera to record the inspection. In addition, the licensee conducted a volumetric ultrasonic examination (UT) and eddy current examination (56 penetrations) from the inside surface (ID) of the penetration base material two inches above and below the J-groove weld.
(2) Identification of boric acid deposits and characterization of deposits	The licensee did not identify any crack-like indications in the BMI penetrations or any evidence of boric acid leakage boric acid deposits on the RPV lower head surface. The licensee observed several boron trails on the RPV lower head surface. The licensee indicated that the residue originated from the previous reactor cavity seal leakage.
(3) Chemical analysis, number of samples of the deposit, and results	In order to establish the source of the boron trails on the RPV lower head, the licensee performed isotopic analysis and energy dispersive spectroscopy or inductively coupled plasma analysis of the smears that were taken from the RPV lower head. Based on the results, the licensee concluded that the stains did not originate from the RCS leakage.
(4) Licensee's actions during outage to clean boric acid deposits from the vessel lower head to establish a baseline for future inspections	The licensee cleaned and inspected the lower head so that, a baseline is established for future inspections.

Table 5.3-4, Inspection Results, Fall 2004 Outage

Inspection Attributes\Plant	D.C. Cook, Unit 2
(1) Type and extent of inspection	The licensee visually inspected all 58 BMI penetrations including 360^0 of each penetration using a remote camera to record the inspection. The visual examination was performed by VT-2 Level II qualified inspectors.
(2) Identification of boric acid deposits and characterization of deposits	The licensee did not identify any boric acid deposits in the annulus region or on the RPV lower head surface. The licensee stated that the RPV lower head condition remained the same as the condition that was left after cleaning it during the previous outage. Based on this observation, the licensee concluded that there was no RCS leakage in the BMI penetration and RPV lower head.
(3) Chemical analysis, number of samples of the deposit, and results	The licensee did not perform chemical analysis.
(4) Licensee's actions during outage to clean boric acid deposits from the vessel lower head to establish a baseline for future inspections	Not addressed.

Table 5.3-4, Inspection Results, Fall 2004 Outage

Inspection Attributes\Plant	Diablo Canyon, Unit 2
(1) Type and extent of inspection	The licensee visually inspected all 58 BMI penetrations including 360^0 of each penetration using a remote camera to record the inspection. The visual examination was performed by VT-2 Level II qualified inspectors.
(2) Identification of boric acid deposits and characterization of deposits	The licensee did not identify any boric acid deposits in the annulus region between the penetration and the reactor vessel. The licensee observed several boron trails with no discernable thickness on the RPV lower head surface. The licensee indicated that the trails originated from the previous reactor cavity seal leakage. The licensee did not identify any degradation of the RPV lower head.
(3) Chemical analysis, number of samples of the deposit, and results	The licensee did not perform chemical analysis.
(4) Licensee's actions during outage to clean boric acid deposits from the vessel lower head to establish a baseline for future inspections	Not addressed.

Table 5.3-4, Inspection Results, Fall 2004 Outage

Inspection Attributes\Plant	Indian Point, Unit 2
(1) Type and extent of inspection	The licensee visually inspected all 58 BMI penetrations including 360^0 of each penetration annulus region using a remote video equipment. The visual examination was performed by VT-2 Level II qualified inspectors. The configuration of RPV lower head insulation did not permit complete examination of the RPV lower head surface.
(2) Identification of boric acid deposits and characterization of deposits	The licensee identified streaks of boron residue with no discernable thickness in the annulus region of most penetrations. Based on the appearance of these streaks and their aging analyses, the licensee concluded that these streaks originated from the previous reactor cavity seal leakage.
(3) Chemical analysis, number of samples of the deposit, and results	In order to establish the source of the boron streaks in the annulus region of the penetrations, the licensee performed aging analysis of the streaks. Based on the results, the licensee concluded that the streaks did not originate from the RCS leakage.
(4) Licensee's actions during outage to clean boric acid deposits from the vessel lower head to establish a baseline for future inspections	The licensee attempted to clean the BMI penetrations but due to equipment problems determined to perform cleaning by lowering the insulation package or by removing the individual panels during the next refuel outage.

Table 5.3-4, Inspection Results, Fall 2004 Outage

Inspection Attributes\Plant	Kewaunee
(1) Type and extent of inspection	The licensee visually inspected all 36 BMI penetrations. The visual examination was witnessed by the Resident Inspector.
(2) Identification of boric acid deposits and characterization of deposits	The licensee identified white streaks and rust colored residue on the RPV lower head. Based on the appearance of these streaks and the chemical analyses, the licensee postulated that these streaks did not originate from the RCS leakage.
(3) Chemical analysis, number of samples of the deposit, and results	In order to establish the source of the boron streaks in the annulus region of the penetrations, the licensee performed chemical analysis of the streaks. Based on the results, the licensee concluded that the streaks did not originate from the RCS leakage.
(4) Licensee's actions during outage to clean boric acid deposits from the vessel lower head to establish a baseline for future inspections	The licensee cleaned and photographed the RPV lower head to establish a baseline for future inspections.

Table 5.3-4, Inspection Results, Fall 2004 Outage

Inspection Attributes\Plant	North Anna, Unit 1
(1) Type and extent of inspection	The licensee visually inspected all 50 BMI penetrations including 360° of each penetration annulus region using mirrors and video equipment using a camera to record the inspection. The visual examination was performed by VT-2 Level II qualified inspectors.
(2) Identification of boric acid deposits and characterization of deposits	The licensee identified white streaks and rust colored residue on the RPV lower head. Based on the appearance of these streaks, the licensee postulated that these streaks did not originate from the RCS leakage. In addition, the licensee identified a tape residue on Inconel to stainless welds of penetrations 36. The residue on the weld of penetration 36 was determined to be acceptable because there was no evidence of boric acid residue on the tape. Presence of a fibrous material on Inconel to stainless welds of penetration 48 was removed and liquid penetrant examination was conducted and no recordable indications were observed on this weld. The licensee, based on its evaluation, concluded that there is no boric acid leakage from the BMI penetrations.
(3) Chemical analysis, number of samples of the deposit, and results	The licensee did not perform chemical analysis.
(4) Licensee's actions during outage to clean boric acid deposits from the vessel lower head to establish a baseline for future inspections	Not addressed.

Table 5.3-4, Inspection Results, Fall 2004 Outage

Inspection Attributes\Plant	Oconee, Unit 3
(1) Type and extent of inspection	The licensee visually inspected 100% of the BMI penetrations using a video camera and direct visual observation.
(2) Identification of boric acid deposits and characterization of deposits	The licensee identified flaking of original coating on the RPV lower head and some superficial corrosion where bare metal was exposed due to loss of coating. The licensee did not identify boric acid residue and wastage of the RPV lower head. Based on these observations, the licensee concluded that there is no RCS leakage.
(3) Chemical analysis, number of samples of the deposit, and results	Not performed.
(4) Licensee's actions during outage to clean boric acid deposits from the vessel lower head to establish a baseline for future inspections	Not addressed.

Table 5.3-4, Inspection Results, Fall 2004 Outage

Inspection Attributes\Plant	Palo Verde, Unit 3
(1) Type and extent of inspection	The licensee visually inspected all 61 BMI penetrations including 360^0 of each penetration using a remote camera. The visual examination was performed by VT-2 inspectors and verified by Level III inspectors.
(2) Identification of boric acid deposits and characterization of deposits	The licensee did not identify any boric acid deposits in the annulus region between the penetration and the reactor vessel. However, small flakes of spraylat (trade name) coating was observed in the annulus region of some of the BMI penetrations. The licensee observed several boron trails with no discernable thickness on the RPV lower head surface. The licensee indicated that the trails originated from the previous reactor cavity seal leakage. The licensee did not identify any through wall leakage or any corrosion of the RPV lower head.
(3) Chemical analysis, number of samples of the deposit, and results	The licensee did not perform chemical analysis.
(4) Licensee's actions during outage to clean boric acid deposits from the vessel lower head to establish a baseline for future inspections	The licensee cleaned and inspected the RPV lower head to establish a baseline for future inspections.

Table 5.3-4, Inspection Results, Fall 2004 Outage

Inspection Attributes\Plant	Prairie Island, Unit 1
(1) Type and extent of inspection	All 36 BMI penetrations including 360^0 of each penetration annulus region of the RPV lower head were examined with a video probe remote camera. The visual examination was performed by VT-2 qualified inspectors.
(2) Identification of boric acid deposits and characterization of deposits	The licensee stated that the general condition of the lower head surface was good. The licensee identified dried liquid streaks which apparently originated from above the BMI penetrations. The licensee postulated that these trails originated from the previous reactor cavity seal. The streaks left traces of thin white deposits in the crevice between majority of the penetrations and the vessel head. Thick deposits were found in the crevice region between penetrations 9,18 and 35 and the vessel head.
(3) Chemical analysis, number of samples of the deposit, and results	The licensee performed radio-chemical and chemical analysis of samples from the penetrations 9,18 and 35, and based on the results the licensee concluded that the deposits did not originate from the RCS leakage.
(4) Licensee's actions during outage to clean boric acid deposits from the vessel lower head to establish a baseline for future inspections	Not addressed.

Table 5.3-4, Inspection Results, Fall 2004 Outage

Inspection Attributes\Plant	Sharon Harris, Unit 1
(1) Type and extent of Inspection	The licensee visually inspected 50 BMI penetrations including 100% of the circumference of each penetration annulus region. The visual examination was performed by VT-2 qualified inspectors. Since the licensee recorded similar inspections during the previous outage, it determined not to record the inspections at this time.
(2) Identification of boric acid deposits and characterization of deposits	The licensee identified thin white streaks on the RPV lower head. Based on its appearance, the licensee concluded that these stains originated from above the RPV lower head. The licensee postulated that these trails originated from the previous reactor cavity seal. Based on these observations, the licensee concluded that there was no RCS leakage on the RPV lower head.
(3) Chemical analysis, number of samples of the deposit, and results	The licensee did not perform chemical analysis.
(4) Licensee's actions during outage to clean boric acid deposits from the vessel lower head prior to restart to establish baseline for future inspections	Not addressed.

Table 5.3-4, Inspection Results, Fall 2004 Outage

Inspection Attributes\Plant	Surry, Unit 1
(1) Type and extent of inspection	The licensee visually inspected all 50 BMI penetrations including 360^0 of each penetration annulus region using mirrors and digital camera to record the inspection. The visual examination was performed by VT-2 Level II qualified inspectors. The licensee inspected 39 BMI penetrations using ultrasonic examination (UT) method with a rotating probe containing two elements in a pitch/catch configuration. This method used the time of flight diffraction (TOFD) technique which provided beam in circumferential and axial directions in the BMI penetration. This examination method was demonstrated to be capable of detecting and sizing flaws in a BMI penetration. In addition, the licensee inspected 48 BMI penetrations using eddy current (ET) examination method.
(2) Identification of boric acid deposits and characterization of deposits	The licensee did not identify any crack-like indications in the BMI penetrations or any evidence of boric acid deposits on the RPV lower head surface. Based on the observation, the licensee postulated that there was no RCS leakage on the RPV lower head.
(3) Chemical analysis, number of samples of the deposit, and results	The licensee did not perform chemical analysis.
(4) Licensee's actions during outage to clean boric acid deposits from the vessel lower head to establish a baseline for future inspections	Even though the licensee did not observe any boric acid deposits on the lower head, it cleaned and inspected the lower head, so that a baseline is established for future inspections.

Table 5.3-4, Inspection Results, Fall 2004 Outage

Inspection Attributes\Plant	Turkey Point, Unit 3
(1) Type and extent of inspection	The licensee inspected all 50 BMI penetrations using UT examination method in lieu of BMV inspection method. The UT examination method used a rotating probe containing two elements in a pitch/catch configuration. This method used the time of flight diffraction (TOFD) technique which provided beam in circumferential and axial directions in the BMI penetration. This examination method was demonstrated to be capable of detecting and sizing flaws in a BMI penetration. In addition, the licensee performed visual inspection of the reactor cavity and underside of the RPV lower head.
(2) Identification of boric acid deposits and characterization of deposits	The licensee did not identify any crack-like indications in the BMI penetrations or any evidence of boric acid deposits on the RPV lower head surface. Based on the observation, the licensee postulated that there was no RCS leakage on the RPV lower head.
(3) Chemical analysis, number of samples of the deposit, and results	The licensee did not perform chemical analysis.
(4) Licensee's actions during outage to clean boric acid deposits from the vessel lower head to establish a baseline for future inspections	Not addressed.

Table 5.3-4, Inspection Results, Fall 2004 Outage

Inspection Attributes\Plant	Indian Point, Unit 3
(1) Type and extent of inspection	The licensee visually inspected all 58 BMI penetrations including 360^0 of each penetration with a remote camera and recorded the inspections. The visual examination was performed by VT-2 inspectors.
(2) Identification of boric acid deposits and characterization of deposits	The licensee identified minor streaks of boron residue with no discernable thickness on 13 BMI penetrations and their annulus region between the penetrations and the RPV lower head. Based on the appearance of these streaks, the licensee noted that they originated from the previous leaks from above the lower head.
(3) Chemical analysis, number of samples of the deposit, and results	The licensee did not perform chemical analysis.
(4) Licensee's actions during outage to clean boric acid deposits from the vessel lower head to establish a baseline for future inspections	The licensee steam cleaned the 13 BMI penetrations which had streaks of boron residue, and reinspected to ensure the effectiveness of steam cleaning of the BMI penetrations and the RPV lower head. The reinspection of the 13 BMI penetrations was performed to establish a baseline for future inspections.

Table 5.3-5, Inspection Results, Spring 2005 Outage

Inspection Attributes\Plant	Palo Verde, Unit 2 (follow-up inspection —original inspection -- fall 2003 outage)
(1) Type and extent of Inspection	The licensee visually inspected all 61 BMI penetrations including 360^0 of each penetration using a remote camera with zoom capabilities to record the inspection. The visual examination was performed by VT-2 inspectors. The licensee cleaned 22 BMI penetrations that were left uncleaned during the previous outage and re-inspected them during this outage.
(2) Identification of boric acid deposits and characterization of deposits	The licensee did not identify any boric acid deposits on any BMI penetrations or on the RPV bottom head surface area. The licensee observed several boron trails with no discernable thickness on the RPV lower head surface. Based on the appearance of these trails, the licensee concluded that these trails did not originate from RCS leakage.
(3) Chemical analysis, number of samples of the deposit, and results	The licensee did not perform chemical analysis.
(4) Licensee's actions during outage to clean boric acid deposits from the vessel lower head to establish a baseline for future inspections	Not addressed.

Table 5.3-5, Inspection Results, Spring 2005 Outage

Inspection Attributes\Plant	South Texas, Unit 1
(1) Type and extent of inspection	The licensee visually inspected all 58 BMI penetrations including entire annulus of each penetration. The visual examination was performed by VT-2 Level II qualified inspectors.
(2) Identification of boric acid deposits and characterization of deposits	The licensee did not identify any boric acid deposits on any BMI penetrations or on the RPV bottom head surface area.
(3) Chemical analysis, number of samples of the deposit, and results	The licensee did not perform chemical analysis.
(4) Licensee's actions during outage to clean boric acid deposits from the vessel lower head to establish a baseline for future inspections	Not addressed.

Table 5.3-5, Inspection Results, Spring 2005 Outage

6 TEMPORARY INSTRUCTION (TI) 2515/152, REVISION 1, REACTOR PRESSURE VESSEL LOWER HEAD PENETRATION NOZZLES (NRC BULLETIN 2003-02)

6.1 INFORMATION REQUESTED BY TI

The NRC developed a temporary instruction to obtain the support of the NRC inspectors in the regions to verify licensee performance of RPV lower head penetration inspections under Bulletin 2003-02 and to document the activities in NRC inspection reports. The NRC inspectors provided qualitative descriptions of the effectiveness of the licensees' examinations. The information requested of the regions by the TI included discussions of the following attributes of the licensee's RPV lower head examinations:

(A)   Type of examination using demonstrated procedures

(B)   Qualifications and training of the personnel performing the examination

(C)   The licensee's ability to disposition and resolve deficiencies

(D)   The licensee's ability to identify pressure boundary leakage as described in the bulletin and/or RPV lower head corrosion

(E)   The physical condition of the RPV lower head (e.g., debris, insulation, dirt, boric acid deposits from other sources, physical layout, viewing obstructions)

(F)   The licensee's ability to identify and characterize small boric acid deposits, and material deficiencies (i.e., cracks, corrosion, etc.) as described in the Bulletin 2003-02.

(G)   Identification of any impediments (e.g., insulation, instrumentation, nozzle distortion) to effective examinations of the RPV lower head

(H)   Licensee's plan to perform appropriate follow-on examinations for indications of boric acid leaks from pressure-retaining components above the RPV lower head

6.2 SUMMARY OF REVIEW OF TI INSPECTION REPORTS

The NRC staff responsible for issuing the bulletin reviewed the reports submitted by the licensee containing their inspection results. A summary of these reviews is contained in Section 5.0. NRC staff also reviewed the NRC inspection report prepared in the regions in response to the TI. The NRC staff's review consisted of an evaluation of the reactor vessel bottom head inspections and verified that the inspections were consistent with the approach recommended by Bulletin 2003-03. Based on the results of the inspections performed by the regions, the NRC staff confirmed that the licensees used qualified personnel, obtained the desired examination coverage, and performed examinations that were capable of finding boric acid deposits of the magnitude discovered at South Texas, Unit 1 in April 2003. Based on the NRC inspections, the NRC staff confirmed the licensees conclusions that there was no evidence of boric acid deposits found during the inspections performed in spring 2003, fall 2003, spring 2004, fall 2004 and spring 2005 outages.

7 INDUSTRY ACTIVITIES RELATED TO BOTTOM MOUNTED NOZZLES

7.1 SUMMARY OF MATERIALS RELIABILITY PROGRAM ACTIVITIES

During the Spring 2003 outage of South Texas Project, Unit 1 (STP-1) discovered leakage on two bottom mounted instrumentation (BMI) penetrations. After performing failure analysis, the licensee concluded that the leakage was due to primary water stress corrosion cracking (PWSCC). The analysis showed that even though the reactor pressure vessel (RPV) lower head temperature is low, the Alloy 600 BMI nozzles are susceptible to PWSCC and will crack under the right conditions. However, the STP-1 experience demonstrated that bare metal visual (BMV) inspection of BMI penetrations is a useful inspection technique for detecting minor leakage and may assist in detecting flaws before they become structurally significant.

As a result of the events at STP-1, the NRC staff issued Bulletin 2003-02, "Leakage from Reactor Pressure Vessel Lower Head Penetrations and Reactor Coolant Pressure Boundary Integrity," on August 21, 2003. The purpose of issuing this bulletin was to advise licensees with pressurized water reactors (PWR) units that current methods of inspecting the RPV lower heads may need to be supplemented with BMV inspections to detect reactor coolant pressure boundary (RCPB) leakage and to request licensees with PWR units to provide the NRC with information related to inspections that had been or would be performed to verify the integrity of the RPV lower head penetrations.

After the issuance of Bulletin 2003-02, the Nuclear Energy Institute (NEI), Material Reliability Program (MRP), and EPRI developed an initiative described in NEI-03-08, "Guideline for the Management of Materials Issues." This initiative consists of a coordinated pro-active program for managing degradation issues and includes addressing the integrity of the BMI penetrations and the RPV) lower head. The MRP is conducting research activities related to the development of inspection techniques for inspecting the BMI penetrations, and is developing a design analysis related to RPV lower head integrity whenever there is a leak in BMI penetrations. The industry developed a strategic plan for detecting aging degradation of the BMI penetrations and the RPV lower head. In a letter dated June 23, 2003, from MRP to the pressurized water reactor owners group (PWROG), MRP recommended that all PWR licensees perform BMV inspections as recommended by the Bulletin 2003-02.

The MRP, thus far, met with the NRC staff on November 25, 2003, July 19, 2004 and September 29, 2005. The purpose of the meetings was to discuss the MRP's strategic plan for detection and control of leakage of the BMI penetrations of the RPV lower head. During these meetings, the MRP discussed the details of integrated industry inspection plan. The main objective of this plan is to develop future inspection methods (described below) and inspection guidelines, review inspection results, develop a model which can be used to assess BMI penetrations' susceptibility to PWSCC and develop inspection guidelines that can be used in monitoring potential degradation of BMI penetrations.

During the September 29, 2005 meeting with the NRC staff, the MRP presented the following long term strategic plan related to the following issues which will be submitted to the NRC staff for review and approval.

- Integrated Industry Plan

  The MRP proposed that the PWR licensees in addition to conducting a baseline BMV inspections, voluntarily perform volumetric inspections of the BMI penetrations. Based on the data, the MRP, in turn will develop a proactive industry aging management program that assures safe and reliable operation.

- NDE Demonstration Program

This program would demonstrate non-destructive examinations (NDE) techniques for use by industry to inspect RPV BMI penetrations. The MRP stated that in addition to BMV inspections of the BMI penetrations, surface inspections [eddy current testing (ECT)], and volumetric inspection of a sample plants, would be conducted over several seasons. For the UT, three types of flaws would be introduced in a test mock-up, specifically Electrical Discharged Machine (EDM) notches, squeezed EDM notches, and cracks.

The MRP indicated that calibration of UT techniques using the test mock-ups with these flaws would enable the licensees to identify any potential flaws in the BMI penetrations. The MRP developed six mock-ups, whereby two mock-ups represented the Westinghouse 2-loop design, two mock-ups represented the Westinghouse 3-loop/4-loop design, and two mock-ups represented the Babcock and Wilcox (B & W) design. The UT techniques detected all flaws in the Westinghouse 3-loop/4-loop design. However, in the Westinghouse 2-loop design 23% of the flaws could not be detected. In the B & W design the UT demonstration of the base metal penetration was not successful due to the repair configuration of the BMI penetrations. The differences between the Westinghouse and the B & W design of the BMI penetrations are shown in Section 2 of this report. So far, the following 12 plants have inspected BMI penetrations using UT. The inspection probes also typically included ECT capabilities. No service-induced cracking has been observed.

Byron Unit 1	McGuire Unit 2
Callaway	Surry Unit 1
Catawba Unit 1	Surry Unit 2
Catawba Unit 2	Turkey Point Unit 3
Diablo Canyon Unit 1	Turkey Point Unit 4
Diablo Canyon Unit 2	Wolf Creek

Previous experience indicated that the UT techniques used for identifying weld defects in the J-groove welds designed by B&W were unsuccessful and the reasons are provided in the following paragraphs:

After the first phase of the Oconee Unit 1 (ONS-1) hot functional test in March 1972, an inspection of the RPV internals revealed several components, including failed BMI penetrations. A visual inspection of the RPV lower head revealed 21 BMI penetrations had broken off. Of the 21 broken BMI penetration, 18 of them failed within 0.125 inch of top of the J-groove weld and the remaining three failed at 0.5 inch above the J-groove weld. It was concluded that the root cause of the failure of the original BMI penetrations at ONS-1 was fatigue caused by flow-induced vibration (FIV).

By the time of the ONS-1 hot functional test in March 1972, all BMI penetrations had already been installed in the RPV lower head of all seven currently operating B&W units. After the ONS-1 hot functional test, the BMI penetration design was modified to strengthen the BMI penetration portion inside the RPV. The portion of the original BMI penetrations inside the RPV were cut off above the J-groove weld, and replaced by a 2-inch OD. Alloy 600 BMI penetration was attached by a full penetration butt weld. Except for Davis-Besse, the modification was performed in the field without post weld heat treatment (PWHT). Figure 2-1 in Section 2 of this document provides details of this modification. The modification for Davis-Besse's BMI penetrations was performed in the shop after the final RPV PWHT.

The modification to the B&W-designed BMI penetrations results in two penetrations joined end to end with two slightly different diameters and potentially a slight diametrical offset. This results in a slight "step" on the BMI penetration ID which causes UT probe lift-off and signal loss when performing the time of flight diffraction (TOFD) UT technique. Current demonstration efforts at the EPRI NDE center have shown this "step" in the BMI penetration to be problematic for obtaining 100% UT coverage and promotes false-positive indications in the B&W BMI penetrations. Industry efforts are currently underway to improve the UT technique for B&W BMI penetrations and with a focus on minimizing the potential for false-positive indications.

- BMI Repairs

The BMI repair part of the plan would define the attributes of an ideal repair, evaluate current repair options versus the ideal attributes, develop new repair technology if necessary, and provide resources for repair technique development as needed.

- Safety Assessment Plan

The MRP's safety assessment plan addressed the following issues:

Materials and fabrication records would be used to validate inspection test mock-ups.

Failure modes and effects analysis (FMEA) of the BMI penetrations.

Deterministic Fracture Mechanics; Loss of Coolant Accident (LOCA) analysis and Probabilistic Risk Assessment

Collateral Damage Assessment

Wastage Evaluation of the RPV Lower Head

The MRP expects to provide the NRC staff additional information from future inspections and aging management program for the BMI penetrations of the RPV lower head in 2007.

7.2 SUMMARY OF ASME CODE ACTIVITIES

In a letter dated August 19, 2002, from Brian Sheron of the NRC to Richard Gimple, ASME Subcommittee on Nuclear Inservice Inspection Chairman, the NRC staff requested that the ASME Section XI Code Committee reevaluate the inspection requirements for all systems that are potentially susceptible to SCC and boric acid corrosion.

In its response to the NRC staff's request, the ASME formed a Task Group on Boric Acid Corrosion to address the inspection requirements associated with boric acid corrosion for all the primary system components in the PWR units. The ASME Section XI, developed a Code Case N-722, "Additional Examinations for PWR Pressure Retaining Welds in Class 1 Components Fabricated with Alloy 600/82/182 Materials ASME Section XI, Division 1," which recommends that BMV inspections be performed every other refueling outage on all the BMI penetrations in the RPV lower head. The industry believes that by implementing the Code Case N-722 the licensees will be able to effectively monitor the aging degradation of the BMI penetrations. The code case contains recommendations for BMV inspections of other alloy 600/82/182 locations in the RCS. The NRC staff has reviewed this code case and is considering incorporating the code case in 10 CFR 50.55a as an augmented inspection requirement.

8 SUMMARY OF FOREIGN EXPERIENCE

The NRC staff requested foreign regulatory agencies provide information regarding PWR inspection results of the BMI penetrations and the RPV lower heads. The NRC staff obtained inspection results from the PWR licensees in Japan, Sweden, France and Belgium. The following summarizes the foreign PWR licensees' inspection methods, inspection results and the frequency of inspections. The NRC staff finds that this information is useful in assessing aging degradation and monitoring of the foreign RPV lower head BMI penetrations.

8.1 INSPECTION OF BMI PENETRATIONS OF PWR UNITS IN JAPAN

As of the date of issuance of this report, only one crack was identified in one BMI penetration at Takahama Unit 1 which is owned by Kansai Electric Power Company (KEPCO). The size of the crack was 0.12 inches (3 mm) long and this crack was identified by eddy current testing (ECT) during the licensee's 21st periodic inspection conducted from November 20, 2002 to March 12, 2003. The licensee suspected that primary water stress corrosion cracking (PWSCC) was the mechanism for the crack initiation in this BMI penetration. The licensee removed the crack during the next inspection period (April 11, 2004 to July 15, 2004) after verifying with ECT that no subsequent crack growth occurred. Based on the worldwide experiences related to PWSCC of the RPV lower head BMI penetrations, National Institute Safety Administration (NISA), the Japanese regulatory agency, issued a regulatory requirement for all the PWR owners to perform BMV inspections of the BMI penetrations every 5 years.

After the issuance of the regulatory requirement, the 22 PWR units out of total 23 units in Japan performed BMV inspections of the RPV lower head BMI penetrations and thus far, no cracking was found. One of the PWR unit has Alloy 690 BMI penetrations which are less susceptible to PWSCC. Out of 9 PWR units, ECT was performed on 8 units, and laser ultrasonic testing (UT) was performed in 1 PWR unit and only one BMI penetration indicated cracking which was subsequently removed as noted above.

8.2 INSPECTION OF BMI PENETRATIONS OF PWR UNITS IN SWEDEN

Information regarding the BMV inspections of the lower head penetrations in RPVs at Sweden indicated that visual inspections were performed on all three PWR units at Ringhals (Units 2, 3 and 4) since 2002. All 50 penetrations at each unit were inspected and so far, no RCS leakage from the BMI penetrations has been found. The inspections were conducted with a remote camera and the results were recorded. No volumetric inspections were performed at these three units and the Swedish licensees plan to continue to perform BMV inspections until 2007.

8.3 INSPECTION OF BMI PENETRATIONS OF PWR UNITS IN FRANCE

In France there are three series of PWR units and each series has a different number of BMI penetrations. BMV inspections were performed by removing the insulation prior to using a remote high resolution video camera and thus far, no RCS leakage was observed in all these penetrations. BMI penetrations in 12 out of 58 RPVs were selected for the UT examinations and these types of examinations will continue until 2008. Some of these BMI penetrations were not subjected to RPV post weld heat treatment (PWHT) operation during construction and some were cold worked after RPV PWHT operation. Thus far, these UT examinations of the BMI penetrations identified no cracking in 18 RPVs however, some non-surface breaking indications were detected which were identified to be fabrication flaws. The French licensees will submit their future inspection schedule for all the BMI penetrations in the PWR fleet to the French Safety Authority for review and approval.

8.4 INSPECTION OF BMI PENETRATIONS OF PWR UNITS IN BELGIUM

Prior to the incident at the STP Unit 1, in years 1995 and 2000, the owner of Doel Unit 1, conducted inspections of the ID of the BMI penetrations using qualified ECT and UT techniques. ECT was used for inner surface examination and UT was used to examine the penetration and penetration-to-weld interface. Small indication i.e., lack of fusion, was observed in some of the 13 inspected BMI penetrations. All indications were reported to be caused by weld geometry effects or lack of fusion/inclusions at the interface between the J-weld and the OD of the BMI penetrations. The penetration on which the largest indication at the weld interface had been identified in 1995 was re-inspected in 2000 but no change was observed.

Following the STP Unit 1 incident, a program of BMV inspection of the BMI penetrations was initiated at the seven Belgium plants starting from September 2003. At the end of 2005, all BMI penetrations at all plants with the exception of the ones in Tihange Unit 1 plant were visually inspected once. At Tihange Unit 1, the insulation is in direct contact with the lower head which made the inspections cumbersome due to limited accessibility. With some modifications, BMV inspections were performed on 16 out of 50 BMI penetrations in 2005 and these inspections resulted in increased radiation exposure to the plant personnel. Thus far, no leakage was found in the BMI penetrations of any PWR units in Belgium.

A schedule for the future BMV inspections of all the PWR BMI penetrations in Belgium has been proposed by the utilities. The first scheduled inspection of the BMI penetrations will be at the Tihange Unit 1 RPV. BMV inspections of all the BMI penetrations at all plants with the exception of the two oldest plants (Doel Units 1 and 2) will be performed every 3 years. The BMI penetrations at Doel Units 1 and 2 will be inspected every 2 years.

8.5 SUMMARY

Based on the BMV inspection results, there is no evidence of active aging degradation of the BMI penetrations at foreign PWR units. Foreign PWR owners, in general, indicated plans to perform additional inspections of BMI penetrations and these inspections are expected to monitor the potential for future degradation of these components.

9   SUMMARY AND CONCLUSIONS

This report presents information regarding the leakage of the bottom mounted instrumentation (BMI) penetrations in the reactor pressure vessel (RPV) lower head which occurred at South Texas Project, Unit 1 (STP-1) during the spring 2003 outage. As a result of this leakage, the NRC staff issued Bulletin 2003-02, "Leakage from Reactor Pressure Vessel Lower Head Penetrations and Reactor Coolant Pressure Boundary Integrity," on August 21, 2003. The purpose of this bulletin was to advise licensees with pressurized water reactors (PWR) units that current methods of inspecting the RPV lower heads may need to be supplemented with bare metal visual (BMV) inspections to detect reactor coolant pressure boundary (RCPB) leakage and to request licensees with PWR units to verify the integrity of the BMI penetrations.

The NRC staff received the inspection plans from all 58 PWR units affected by the bulletin and this information has been included in this report. The responses included licensees' proposal to perform BMV inspection of the BMI penetrations in upcoming outages, their commitment to future inspections beyond the upcoming inspections of the RPV lower head and its penetrations, and their plans to clean the RPV lower head to establish baseline criteria for future inspections.

The NRC staff has received the inspection results from all 58 PWR units. The BMV inspections of the RPV lower head penetration were performed by 3 units during spring 2003 outage (prior to the issuance of the Bulletin 2003-02), 23 units during the fall 2003 outage, 16 units during the spring 2004 outage, 14 units during the fall 2004 outage, and 2 units during the spring 2005 outage. The NRC staff received the inspection results from the licensees which included the type and extent of inspections, identification and characterization of boric acid deposits, and each licensee's action to clean boric acid deposits from the RPV lower head to establish a baseline for future inspections.

All licensees required to respond to the bulletin provided a commitment to perform the BMV inspections recommended by the bulletin or had already performed a recent BMV inspection of their BMIs. All licensees required to respond provided a summary of their inspection results and the NRC verified that these inspections were completed as part of the NRC regional inspection program. So far, no evidence of leakage has been found in the BMI penetrations of RPV lower heads.

Most licensees indicated in their initial response that they intended to perform subsequent inspections of their BMIs based on either a specific schedule, such as once every third refueling outage, or a schedule recommended by industry or the NRC. Some licensees did not indicate in their responses any plans to perform subsequent inspections. Since regulatory requirements to perform BMV inspections on a specific schedule do not yet exist, the NRC staff did not pursue this matter in follow-up communications. However, both the MRP and the NRC are pursuing the development of long term inspection programs to ensure that any further degradation in BMIs is detected and corrected.

The industry, which is represented by Nuclear Energy Institute (NEI), Material Reliability Program (MRP) and EPRI, proposed to implement an integrated industry inspection plan for the BMI penetrations. This plan, in addition to BMV inspections, may include volumetric inspections of the BMI penetrations to monitor their aging degradation. The industry also proposed to develop a new repair technology, if necessary, and provide resources for repair technique development for the BMI penetrations. In addition, the ASME Code developed a Code Case N-722, "Additional Examinations for PWR Pressure Retaining Welds in Class 1 Components Fabricated with Alloy 600/82/182 Materials ASME Section XI, Division 1," which recommends that BMV inspections be performed every other refueling outage on all the BMI penetrations in the RPV lower head.

This report provides a brief summary of BMV inspections of the BMI penetrations that were performed by foreign licensees, the inspection results, and the foreign licensee's future inspections plans.

In conclusion, this report summarizes the available information regarding the current and future inspection programs and industry's and NRC's on-going plans to ensure the integrity of the BMI penetrations. Even though there was no evidence of BMI penetration leakage at the foreign PWR units and at the 58 PWR units in the US, the NRC staff believes that continuous monitoring of the BMI penetrations is necessary to ensure the integrity of the RPV lower head.

APPENDIX A

OMB Control No.: 3150-0012   August 21, 2003

UNITED STATES
NUCLEAR REGULATORY COMMISSION
OFFICE OF NUCLEAR REACTOR REGULATION
WASHINGTON, DC 20555

NRC BULLETIN 2003-02:   LEAKAGE FROM REACTOR PRESSURE VESSEL LOWER HEAD
                          PENETRATIONS AND REACTOR COOLANT PRESSURE BOUNDARY
                          INTEGRITY

Addressees

All holders of operating licenses for pressurized-water nuclear power reactors (PWRs) with
penetrations in the lower head of the reactor pressure vessel (RPV), except those who have
permanently ceased operations and have certified that fuel has been permanently removed from the
reactor pressure vessel.

All other holders of operating licenses for nuclear power plants will receive a copy of this bulletin for
information.

Purpose

The US Nuclear Regulatory Commission (NRC) is issuing this bulletin to:

(1) advise PWR addressees that current methods of inspecting the RPV lower heads may need to be
     supplemented with additional measures (e.g., bare-metal visual inspections) to detect reactor
     coolant pressure boundary (RCPB) leakage,

(2) request PWR addressees to provide the NRC with information related to inspections that have
     been or will be performed to verify the integrity of the RPV lower head penetrations, and

(3) require PWR addresses to provide a written response to the NRC in accordance with the
     provisions of Section 50.54(f) of Title 10 of the *Code of Federal Regulations* (10 CFR 50.54(f)).

Background

PWR RPV upper heads have a number of penetrations, including penetrations for control rod drive
mechanisms (CRDMs). These penetrations are typically made of nickel-based Inconel Alloy 600. The
penetrations are welded to the inside of the RPV head with nickel-based Inconel Alloy 82/182
materials. Most PWRs also have penetrations in the RPV lower heads for in-core nuclear
instrumentation. The same Inconel materials are typically used in the lower head penetrations and
welds. The primary coolant water and the operating conditions of PWR plants have caused cracking of

nickel-based alloys in upper head penetrations through a process called primary water stress corrosion cracking (PWSCC).

As part of the response to issues associated with degradation of the RPV upper head at the Davis-Besse Nuclear Power Station, the NRC issued Bulletin 2002-01, "Reactor Pressure Vessel Head Degradation and Reactor Coolant Pressure Boundary Integrity," dated March 18, 2002. This bulletin requested information about the condition and inspections of RPV upper heads and about licensee's boric acid corrosion control (BACC) programs. The NRC subsequently issued Bulletin 2002-02, "Reactor Pressure Vessel Head and Vessel Head Penetration Nozzle Inspection Programs," dated August 9, 2002. This bulletin was issued to address NRC staff concerns regarding the adequacy of visual examinations as a primary inspection method for the RPV upper head and RPV upper head penetrations. By NRC Order EA-03-009, dated February 11, 2003, the NRC required specific inspections of RPV upper heads, CRDM penetrations, and associated welds in addition to the inspections required by Section XI of the American Society of Mechanical Engineers (ASME) Boiler and Pressure Vessel Code (Code).

After evaluating the responses received in response to Bulletin 2002-01, the NRC staff issued requests for additional information (RAIs) to PWR licensees in order to obtain more detailed information regarding licensee BACC programs. The NRC staff summarized its review of the responses to Bulletin 2002-01 and the associated RAIs in Regulatory Issue Summary (RIS) 2003-13, "NRC Review of Responses to Bulletin 2002-01, 'Reactor Pressure Vessel Head Degradation and Reactor Coolant Pressure Boundary Integrity,'" dated July 29, 2003. The NRC noted in RIS 2003-13 that most licensees do not perform inspections of Alloy 600/82/182 materials beyond those required by Section XI of the ASME Code to identify potential cracked and leaking components. For the RPV lower head, the ASME Code specifies that a visual examination, called a VT-2 examination, be performed during system pressure testing. Licensees may meet the ASME Code requirement for a VT-2 inspection by performing an inspection of the RPV lower head without removing insulation from around the head and penetrations. It is the NRC staff's understanding that many licensees perform the ASME Code-required inspections without removing insulation and, therefore, may not be able to detect the amounts of through-wall leakage expected from potential flaws due to PWSCC or other cracking mechanisms.

The lower head and bottom mounted instrumentation (BMI) penetrations of the South Texas Project Unit 1 (STP Unit 1) RPV were visually inspected on April 12, 2003, as a routine part of the unit's refueling outage. The lower head of the reactor is surrounded by an insulating box structure with no insulation directly in contact with the lower head. The inspection was accomplished by removing three of the insulation panels forming the insulating box. Three different vantage points were used to inspect all 58 BMI penetrations in the vessel lower head. The inspection found small amounts of white residue around two of the 58 BMI penetrations (numbers 1 and 46) at the junction where the penetrations met the lower reactor vessel head. The residue at penetrations 1 and 46 was collected for laboratory analysis to determine the source of the residue material. Approximately 150 milligrams and 3 milligrams were collected from penetrations 1 and 46, respectively. The analysis of the sample for lithium demonstrated that the lithium was approximately 99.9 percent lithium-7, which indicated that the reactor coolant system was the source of the residue. The analysis of the sample for cesium indicated that the average age of the residue collected was between 3 and 5 years. The licensee for STP Unit 1 indicated that these residues were not visible during the previous inspection on November 20, 2002.

Ultrasonic inspections (using circumferential, axial, and zero degree probes) of 57 BMI penetration tubes at STP Unit 1 were completed in May 2003, along with the visual inspections of the surfaces of the 58 J-groove welds which attach the BMI penetration tubes to the RPV lower head. In addition, eddy current testing (ECT) was used to examine the J-groove weld and inside diameter surfaces of some BMI penetration tubes. Axial cracks were found in penetration tubes 1 and 46. The largest of these cracks was entirely through-wall and extended above and below the J-groove weld. No evidence of cracking was found in any other penetration. BMI penetrations 1 and 46 have been repaired. The licensee is continuing to investigate the cause of the cracks. The investigation has not, to date, identified any manufacturing practice or operating condition that is unique to the affected penetrations or to the RPV at STP Unit 1. The design of the area beneath the RPV at STP Unit 1 and the inspection methods used by the licensee enabled the discovery of the leaking penetrations. From the NRC staff reviews described in RIS 2003-13, the NRC staff concluded that leakage such as that observed at STP Unit 1 would likely not have been detected during inspections performed at many other PWRs.

Discussion

The RPV and its head penetrations are an integral part of the RCPB, and their integrity is important to the safe operation of the plant. The recent identification of cracking and leakage from two BMI penetrations at STP Unit 1 raises questions about potential degradation mechanisms which may be active in this area. In addition, licensee responses to the Bulletin 2002-01 followup RAIs raised questions about the adequacy of inspections performed by licensees to detect leakage from RPV lower head penetrations.

As indicated above, the investigation of the degradation mechanism involved in the cracking of the two penetrations at STP Unit 1 is continuing. However, an evaluation of the available information leads to several observations. First, although the root cause of the cracking experienced at STP Unit 1 is not yet understood, the investigation to date has not identified potential root causes which would be unique to the affected penetrations at STP Unit 1.

Second, the licensee for STP Unit 1 uses a method of inspecting the RPV lower head penetrations that permits visual examination of the external metal surfaces of the vessel lower head and its penetrations, unimpeded by the surrounding insulation. In comparison to the previously discussed VT-2 examinations specified in Section XI of the ASME Code, which do not require the removal of insulation and must be performed at normal operating pressure conditions once each refueling outage, the inspections conducted by the STP Unit 1 licensee are superior for the purpose of finding evidence of leakage like that observed at STP Unit 1. In fact, the NRC staff has concluded that the VT-2 examinations required by Section XI of the ASME Code would not be effective at finding deposits like those discovered at STP Unit 1.

Third, the circumstances of the STP Unit 1 findings indicate that the cracking and the onset of leakage may have occurred several years prior to the discovery of leakage. The licensee's prior inspections of STP Unit 1 lower head were capable of finding the deposits observed in April 2003. However, no evidence of leakage had been noted as the result of any inspections conducted prior to April 2003. Therefore, a one-time inspection of an RPV lower head area may not provide adequate assurance that degradation is not occurring similar to that observed in the BMI penetrations at STP Unit 1.

The small amount of leakage from the cracks discovered at STP Unit 1 did not represent an immediate safety problem due to the size and orientation of the cracks. In addition, safety systems included in plant designs and required to be available during plant operation would be able to mitigate the effects of more significant leaks, including a gross rupture of an RPV lower head penetration. Although unlikely, a significant leak from an RPV lower head penetration could introduce operational and safety concerns since it would require operation of safety systems for an extended period and complicate longer term efforts to stabilize the plant. To maintain the overall defense-in-depth philosophy incorporated into the design and operation of nuclear power plants, licensees should take appropriate actions to ensure the integrity of the RPV lower head penetrations.

The NRC staff believes it is appropriate for licensees to assess their current inspection practices to periodically ensure that there are no leaks from RPV lower head penetrations. This conclusion is based on the safety concerns associated with a significant leak from the RPV lower head and the uncertainties associated with the ability of some current inspection practices to identify cracks and resultant small leaks from RPV lower head penetrations.

Inspections capable of detecting through-wall leakage from any RPV lower head penetration, beginning at the next refueling outage, would provide additional confidence in the integrity of the RPV lower head penetrations. If visual inspections are performed to detect evidence of possible leakage, such inspections should include an inspection of 100% of the circumference of each penetration as it enters the RPV lower head.

The industry's Materials Reliability Program (MRP) has made recommendations for PWR licensees to perform bare-metal visual inspections of RPV lower head penetrations during the current or next refueling outage. The recommendations were included in a letter from Leslie Hartz, MRP Senior Representative, dated June 23, 2003 (Agencywide Documents Access and Management System (ADAMS) Accession No. ML031920395). The MRP is an industry program, coordinated by EPRI, to address material-related issues associated with PWRs.

The NRC is aware that preexisting conditions at some facilities may prevent licensees from performing bare-metal visual inspections of some RPV lower head penetrations during their next refueling outage. For these plants, such inspections of the RPV lower head penetrations may not be possible, for example, until after plant modifications, cleaning, and completion of other tasks provide access and a clean surface for baseline and future inspections. For the plants unable to perform inspections as recommended above, additional confidence in the integrity of the RPV lower head penetrations may be obtained by licensees (1) developing an inspection plan to examine as many of the RPV lower head penetrations as is practical, and (2) taking the necessary steps to enable the performance of inspections as above for each penetration during subsequent refueling outages. In conducting inspections or other activities on the RPV lower head, licensees should recognize that entry into and work in cavities under PWR reactor vessels present very high radiation hazards. Access controls to these areas should require, among other things, close communication between plant operations and radiation protection staff on the status of the highly activated components (e.g., thimble retraction from the core into the reactor cavity) so that required reactor cavity access controls and oversight can be fully implemented before very high radiation levels are created. More information on these under-vessel hazards is provided in Appendix B of Regulatory Guide 8.38, "Control Of Access To High And Very High Radiation Areas In Nuclear Power Plants."

The NRC staff is working with the industry and other stakeholders to revise the ASME Code and NRC regulations to address inspection of RCPB locations susceptible to cracking, including RPV penetrations. These activities will not be completed for several years, so the NRC is issuing this bulletin to address the immediate concerns identified following the reviews of the responses to Bulletin 2002-01 and followup RAIs and the discovery of leaks from BMI penetrations at STP Unit 1. The NRC has posted and will continue to post information about these subjects on its Web site (www.nrc.gov).

Applicable Regulatory Requirements

The NRC has acknowledged that the existing regulatory requirements may need to be supplemented in order to ensure required inspections of RPV lower head penetrations are adequate to identify potential penetration leakage. However, several provisions of the NRC regulations and plant operating licenses (technical specifications) pertain to RCPB integrity and the issues addressed by this bulletin. The general design criteria (GDC) for nuclear power plants (Appendix A to 10 CFR Part 50), or, as appropriate, similar requirements in the licensing basis for a reactor facility, the requirements of 10 CFR 50.55a, and the quality assurance criteria of Appendix B to 10 CFR Part 50 provide the bases and requirements for NRC staff assessment of the potential for, and consequences of, degradation of the RCPB.

The applicable GDCs include GDC 14 (Reactor Coolant Pressure Boundary), GDC 31 (Fracture Prevention of Reactor Coolant Pressure Boundary), and GDC 32 (Inspection of Reactor Coolant Pressure Boundary). GDC 14 specifies that the RCPB be designed, fabricated, erected, and tested so as to have an extremely low probability of abnormal leakage, of rapidly propagating failure, and of gross rupture. GDC 31 specifies that the probability of rapidly propagating fracture of the RCPB be minimized. GDC 32 specifies that components which are part of the RCPB have the capability of being periodically inspected to assess their structural and leaktight integrity.

NRC regulations in 10 CFR 50.55a state that ASME Class 1 components (which includes the RCPB) must meet the requirements of Section XI of the ASME Code. Various portions of the ASME Code address RCPB inspection. For example, Table IWB-2500-1 of Section XI of the ASME Code provides examination requirements during system leakage testing of all pressure-retaining components of the RCPB and references IWB-3522 for acceptance standards. IWB-3522.1©) and (e) specify that conditions requiring correction include the detection of leakage from insulated components and discoloration or accumulated residues on the surfaces of components, insulation, or floor areas that may be evidence of borated water leakage, with leakage defined as the through-wall leakage that penetrates the pressure retaining membrane. Therefore, 10 CFR 50.55a, by reference to the ASME Code, does not permit through-wall degradation of the RPV lower head penetrations. For through-wall leakage identified by visual examinations in accordance with the ASME Code, acceptance standards for the identified degradation are provided in IWB-3142. Specifically, supplemental examination (by surface or volumetric examination), corrective measures or repairs, analytical evaluation, and replacement provide methods for determining the acceptability of degraded components.
Criterion V (Instructions, Procedures, and Drawings) of Appendix B to 10 CFR Part 50 states that activities affecting quality shall be prescribed by documented instructions, procedures, or drawings of a type appropriate to the circumstances and shall be accomplished in accordance with these instructions, procedures, or drawings. Criterion V further states that instructions, procedures, or drawings shall include appropriate quantitative or qualitative acceptance criteria for determining that important activities have been satisfactorily accomplished. Visual and volumetric examinations of the RCPB are activities that should be documented in accordance with these requirements.

Criterion IX (Control of Special Processes) of Appendix B to 10 CFR Part 50 states that special processes, including nondestructive testing, shall be controlled and accomplished by qualified personnel using qualified procedures in accordance with applicable codes, standards, specifications, criteria, and other special requirements.

Criterion XVI (Corrective Action) of Appendix B to 10 CFR Part 50 states that measures shall be established to assure that conditions adverse to quality are promptly identified and corrected. For significant conditions adverse to quality, the measures taken shall include root cause determination and corrective action to preclude repetition of the adverse conditions. For degradation of the RCPB, the root cause determination is important for understanding the nature of the degradation present and the required actions to mitigate future degradation. These actions could include proactive inspections and repair of degraded portions of the RCPB.

Plant technical specifications (TS) pertain to this issue insofar as they do not allow operation with through-wall reactor coolant system pressure boundary leakage.

Requested Information

(1) All subject PWR addressees are requested to provide the following information. The responses for facilities that will enter refueling outages before December 31, 2003, should be provided within 30 days of the date of this bulletin. All other responses should be provided within 90 days of the date of this bulletin.

(a) A description of the RPV lower head penetration inspection program that has been implemented at your plant. The description should include when the inspections were performed, the extent of the inspections with respect to the areas and penetrations inspected, inspection methods used, the process used to resolve the source of findings of any boric acid deposits, the quality of the documentation of the inspections (e.g., written report, video record, photographs), and the basis for concluding that your plant satisfies applicable regulatory requirements related to the integrity of the RPV lower head penetrations.

(b) A description of the RPV lower head penetration inspection program that will be implemented at your plant during the next and subsequent refueling outages. The description should include the extent of the inspections which will be conducted with respect to the areas and penetrations to be inspected, inspection methods to be used, qualification standards for the inspection methods, the process used to resolve the source of findings of boric acid deposits or corrosion, the inspection documentation to be generated, and the basis for concluding that your plant will satisfy applicable regulatory requirements related to the structural and leakage integrity of the RPV lower head penetrations.

(c)   If you are unable to perform a bare-metal visual inspection of each penetration during the next refueling outage because of the inability to perform the necessary planning, engineering, procurement of materials, and implementation, are you planning to perform bare-metal visual inspections during subsequent refueling outages? If so, provide a description of the actions that are planned to enable a bare-metal visual inspection of each penetration during subsequent refueling outages. Also, provide a description of any penetration inspections you plan to perform during the next refueling outage. The description should address the applicable items in paragraph (b).

(d)   If you do not plan to perform either a bare-metal visual inspection or non-visual (e.g., volumetric or surface) examination of the RPV lower head penetrations at the next or subsequent refueling outages, provide the basis for concluding that the inspections performed will assure applicable regulatory requirements are and will continue to be met.

(2) Within 60 days of plant restart following the next inspection of the RPV lower head penetrations, the subject PWR addressees should submit to the NRC a summary of the inspections performed, the extent of the inspections, the methods used, a description of the as-found condition of the lower head, any findings of relevant indications of through-wall leakage, and a summary of the disposition of any findings of boric acid deposits and any corrective actions taken as a result of indications found.

Required Response

In accordance with 10 CFR 50.54(f), the subject PWR addressees are required to submit written responses to this bulletin. This information is sought to verify licensees' compliance with the current licensing basis for the subject PWR addressees. The addressees have two options:

(1) addressees may choose to submit written responses providing the information requested above within the requested time periods, or

(2) addressees who choose not to provide the information requested or cannot meet the requested completion dates are required to submit written responses within 15 days of the date of this bulletin. The responses must address any alternative course of action proposed, including the basis for the acceptability of the proposed alternative course of action.

The required written responses should be addressed to the US Nuclear Regulatory Commission, ATTN: Document Control Desk, 11555 Rockville Pike, Rockville, Maryland 20852, under oath or affirmation under the provisions of Section 182a of the Atomic Energy Act of 1954, as amended, and 10 CFR 50.54(f). In addition, a copy of a response should be submitted to the appropriate regional administrator.

Reasons for Information Request

NRC regulatory requirements and plant TS requirements preclude operation with through-wall leakage from the RCPB. Requirements in the ASME Code, NRC regulations, and plant TS are intended to make licensees perform inspections to maintain an extremely low probability of abnormal leakage, of rapidly propagating failure, and of gross rupture.

The current inspection techniques used at many PWRs may not detect small leaks such as those discovered at STP Unit 1. Uncertainty exists about the root cause of the cracking and resultant leakage at STP Unit 1, and whether other PWRs with RPV lower head penetrations could have similar problems. A detailed assessment of the risks associated with this issue is hampered by the uncertainties associated with the degradation mechanisms which may be active in RPV lower head penetrations, plant conditions (especially for those plants that have not performed the recommended inspections), and the course of events given a significant leak from the lower head. Improved inspections of the RPV lower head penetrations will resolve some of these uncertainties and could identify and allow correction of conditions before they become a significant safety concern.

This information request is necessary to permit the NRC staff to verify compliance with existing regulations and plant-specific licensing bases. The information being requested by this bulletin focuses on RPV lower head penetrations in more detail than previous generic communications and, therefore, is not currently available to the NRC staff. The NRC staff will use the information to assess the acceptability of current licensee lower vessel head inspection programs to identify BMI penetration leakage, and to determine the need for, and guide the development of, any additional regulatory actions (e.g., generic communications, orders, or rulemaking) to address the integrity of the RCPB. Such regulatory actions could include regulatory requirements for augmented inspection programs under 10 CFR 50.55a(g)(6)(ii). The NRC staff will review the responses to this bulletin to determine whether the PWR addressees' inspections provide reasonable assurance that existing applicable regulations are met. If concerns are identified, the NRC staff will contact each affected addressee.

Related Generic Communications

Regulatory Issue Summary 2003-13, "NRC Review of Responses to Bulletin 2002-01, 'Reactor Pressure Vessel Head Degradation and Reactor Coolant Pressure Boundary Integrity,' July 29, 2003 (ADAMS Accession No. ML032100653)

Information Notice 2003-11 "Leakage Found on Bottom-Mounted Instrumentation Nozzles," August 13, 2003 (ADAMS Accession No. ML032250135)

Bulletin 2002-02, "Reactor Pressure Vessel Head and Vessel Head Penetration Nozzle Inspection Programs," August 9, 2002 (ADAMS Accession No. ML022200494)

Bulletin 2002-01, "Reactor Pressure Vessel Head Degradation and Reactor Coolant Pressure Boundary Integrity," March 18, 2002 (ADAMS Accession No. ML020770497)

Generic Letter 88-05, "Boric Acid Corrosion of Carbon Steel Reactor Pressure Boundary Components in PWR Plants," March 17, 1988 (ADAMS Accession No. ML031130424)

Backfit Discussion

Under the provisions of Section 182a of the Atomic Energy Act of 1954, as amended, and 10 CFR 50.54(f), this bulletin transmits an information request for the purpose of verifying compliance with existing applicable regulatory requirements (see the Applicable Regulatory Requirements section of this bulletin). Specifically, the required information will enable the NRC staff to determine whether current inspection and maintenance practices for the detection of degradation of the RCPB at reactor facilities (similar to the degradation observed at STP Unit 1) provide reasonable assurance that RCPB integrity is being maintained. No backfit is either intended or approved by the issuance of this bulletin, and the staff has not performed a backfit analysis.

Federal Register Notification

A notice of opportunity for public comment on this bulletin was not published in the *Federal Register* because the NRC staff is requesting information from power reactor licensees on an expedited basis for the purpose of assessing compliance with existing applicable regulatory requirements and the need for subsequent regulatory action. This bulletin was prompted by the discovery of leaks from BMI penetrations at STP Unit 1 and by the NRC staff's assessment of responses to Bulletin 2002-01. As the resolution of this matter progresses, the opportunity for public involvement will be provided. Nevertheless, comments on the actions requested and the technical issues addressed by this bulletin may be sent to the US Nuclear Regulatory Commission, ATTN: Document Control Desk, Washington, DC 20555-0001.

Small Business Regulatory Enforcement Fairness Act

The NRC has determined that this action is not subject to the Small Business Regulatory Enforcement Fairness Act of 1996.

Paperwork Reduction Act Statement

This bulletin contains an information collection that is subject to the Paperwork Reduction Act of 1995 (44 U.S.C. 3501 et seq.). This information collection was approved by the Office of Management and Budget, clearance no. 3150-0012, which expires August 31, 2006. The burden to the public for this mandatory information collection is estimated to average 110 hours per response, including the time for reviewing instructions, searching existing data sources, gathering and maintaining the data needed, and completing and reviewing the information collection. Send comments regarding this burden estimate or any other aspect of this information collection, including suggestions for reducing the burden, to the Records Management Branch (T-6 E6), US Nuclear Regulatory Commission, Washington, DC 20555-0001, or by Internet electronic mail to INFOCOLLECTS@NRC.GOV; and to the Desk Officer, Office of Information and Regulatory Affairs, NEOB-10202, (3150-0012), Office of Management and Budget, Washington, DC 20503.

Public Protection Notification

The NRC may not conduct or sponsor, and a person is not required to respond to, an information collection unless the requesting document displays a currently valid OMB control number.

If you have any questions about this matter, please contact one of the persons listed below or the appropriate Office of Nuclear Reactor Regulation project manager.

/RA/
Bruce A. Boger, Director
Division of Inspection Program Management
Office of Nuclear Reactor Regulation

Technical Contact:	Edmund Sullivan
	301-415-2796
	E-mail: ejs@nrc.gov
Lead Project Manager:	Stephen R. Monarque
	301-415-1544
	E-mail: srm2@nrc.gov

APPENDIX B

REFERENCES

NRC Generic Communication

1. NRC Bulletin 2003-02, "Leakage from Reactor Pressure Vessel Lower Head Penetrations and Reactor Coolant Pressure Boundary Integrity," US Nuclear Regulatory Commission, Washington, DC, ADAMS Accession No. ML032320153, August 21, 2003.

Licensees' 30-day and 90-day Responses to NRC Staff

1. Cotton, Sherrie R. Response to NRC Bulletin 2003-02 Regarding Reactor Vessel Lower Head Nozzle Integrity, Arkansas Nuclear One, Unit 1 Docket No. 50-313 License No. DPR-51, ADAMS Accession No. ML033250535, November 19, 2003.

2. Pearce, William L. *Beaver Valley Power Station, Unit No. 1 and No. 2, BV-1 Docket No. 50-334, License No. DPR-66, BV-2 Docket No. 50-412, License No. NPF-73, Response to NRC Bulletin 2003-02,* ADAMS Accession No. ML032680872, September 19, 2003.

3. Jury, Keith R. *Thirty-Day Response to NRC Bulletin 2003-02, "Leakage from Reactor Pressure Vessel Lower Head Penetrations and Reactor Coolant Pressure Boundary Integrity,"* (Braidwood Station, Units 1 and 2; Byron Station, Units 1 and 2; and Three Mile Island, Unit 1), ADAMS Accession No. ML032691399, September 22, 2003.

4. Young, Keith D. *Response to NRC Bulletin 2003-02, "Leakage from Reactor Pressure Vessel Lower Head Penetrations and Reactor Coolant Pressure Boundary Integrity,"* (Callaway Plant, Unit 1), ADAMS Accession No. ML033381051, November 19, 2003.

5. McCollum, W.R. *Response to NRC Bulletin 2003-02 "Leakage from Reactor Pressure Vessel Lower Head Penetrations and Reactor Coolant Pressure Boundary Integrity,"* (McGuire and Catawba Nuclear Stations, Units 1 and 2; and Oconee Nuclear Station, Units 1, 2, and 3), ADAMS Accession No. ML032660390, September 18, 2003.

6. Powers, R.P. *Nuclear Regulatory Commission Bulletin 2003-02: Leakage from Reactor Pressure Vessel Lower Head Penetrations and Reactor Coolant Pressure Boundary Integrity Thirty-Day Response,* (Donald C. Cook Nuclear Plant, Units 1 and 2), ADAMS Accession No. ML032690949, September 17, 2003.

7. Walker, Roger D. *Comanche Peak Steam Electric Station (CPES) Docket Nos. 50-445 and 50-446 Response to NRC Bulletin 2003-02, "Leakage from Reactor Pressure Vessel Lower Head Penetrations and Reactor Coolant Pressure Boundary Integrity,"* (Comanche Peak Steam Electric Station, Units 1 and 2), ADAMS Accession No. ML032730595, September 19, 2003.

8. Young, Dale E. *Crystal River, Unit 3 - 30-Day Response to Bulletin 2003-02, "Leakage from Reactor Pressure Vessel Lower Head Penetrations and Reactor Coolant Pressure Boundary Integrity,"* ADAMS Accession No. ML032661246, September 17, 2003.

9. Myers, Lew W. *Davis-Besse Nuclear Power Station Response to NRC Bulletin 2003-02, "Leakage from Reactor Pressure Vessel Lower Head Penetrations and Reactor Coolant Pressure Boundary Integrity,"* ADAMS Accession No. ML033250499, November 19, 2003.

10. Womack, Lawrence F. *90-Day Response to NRC Bulletin 2003-02, "Leakage From Reactor Pressure Vessel Lower Head Penetrations and Reactor Coolant Pressure Boundary Integrity,"* (Diablo Canyon, Units 1 and 2), ADAMS Accession No. ML033370064, November 18, 2003.

11. Gasser, Jeffrey T. *Response to NRC Bulletin 2003-02 "Leakage From Reactor Pressure Vessel Lower Head Penetrations and Reactor Coolant Pressure Boundary Integrity,"* (Joseph M. Farley Nuclear Plant, Units 1 and 2; and Vogtle Electric Generating Plant, Units 1 and 2), ADAMS Accession No. ML032660442, September 19, 2003.

12. Widay, Joseph A. *Response to Bulletin 2003-02, "Leakage from Reactor Pressure Vessel Lower Head Penetrations and Reactor Coolant Pressure Boundary Integrity,"* (R. E. Ginna Nuclear Power Plant, Unit 1), ADAMS Accession No. ML032690163, September 19, 2003.

13. Scarola, James. *90-day Response to NRC Bulletin 2003-02 for "Leakage from Reactor Pressure Vessel Lower Head Penetrations and Reactor Coolant Pressure Boundary Integrity,"* (Shearon Harris Nuclear Power Plant, Unit 1), ADAMS Accession No. ML033280469, November 13, 2003.

14. Dacimo, Fred R. *90-Day Response to NRC Bulletin 2003-02 Regarding Leakage From Reactor Pressure Vessel Lower Head Penetrations and Reactor Coolant Pressure Boundary Integrity,* (Indian Point, Units 2 and 3), ADAMS Accession No. ML033220344, November 13, 2003.

15. Coutu, Thomas. *Nuclear Regulatory Commission Bulletin 2003-02: "Leakage from Reactor Pressure Vessel Lower Head Penetrations and Reactor Coolant Pressure Boundary Integrity - 90-day Response,"* (Kewaunee Nuclear Power Plant, Unit 1), ADAMS Accession No. ML033240427, November 10, 2003.

16. Christian, David A. *Ninety-day Response to NRC Bulletin 2003-02, "Leakage from Reactor Pressure Vessel Lower Head Penetrations And Reactor Coolant Pressure Boundary Integrity,"* (North Anna Power Station, Units 1 and 2; and Millstone Power Station, Unit 3), ADAMS Accession No. ML033280491, November 17, 2003.

17. Nauldin, David. *APS' 30-Day Response to the Information Requested by NRC Bulletin 2003-02,* (Palo Verde Nuclear Generating Station, Units 1, 2, 3), ADAMS Accession No. ML032731333, September 19, 2003.

18. Cayla, A. J. *Nuclear Regulatory Commission Bulletin 2003-02: Leakage from Reactor Pressure Vessel Lower Head Penetrations and Reactor Coolant Pressure Boundary Integrity - 30-Day Response,* (Point Beach Nuclear Plant, Units 1 and 2), ADAMS Accession No. ML032671205, September 22, 2003.

19. Solymossy, Joseph M. *Nuclear Regulatory Commission Bulletin 2003-02: Leakage from Reactor Pressure Vessel Lower Head Penetrations and Reactor Coolant Pressure Boundary Integrity - 30-day Response,* (Prairie Island Nuclear Generating Plant, Units 1 and 2), ADAMS Accession No. ML032690983, September 19, 2003.

20. Lucas, J. F. *Submittal of 90-Day Response to NRC Bulletin 2003-02, "Leakage from Reactor Pressure Vessel Lower Head Penetrations and Reactor Coolant Pressure Boundary Integrity,"* (H. B. Robinson Steam Electric Plant, Unit 2), ADAMS Accession No. ML033220355, November 13, 2003.

21. Carlin, John. *30-Day Response to NRC Bulletin 2003-02, "Leakage from Reactor Pressure Vessel Lower Head Penetrations and Reactor Coolant Pressure Boundary Integrity," Salem Generating Station Units I and 2,* ADAMS Accession No. ML032671216, September 11, 2003.

22. Stall, J. A. *NRC Bulletin 2003-02, "Leakage From Reactor Pressure Vessel Lower Head Penetrations and Reactor Coolant Pressure Boundary Integrity,"* (Turkey Point Units 3 and 4; and Seabrook Station), ADAMS Accession No. ML032671199, September 19, 2003.

23. Burzynski, Mark J. *Sequoyah Nuclear Plant (SQN) Units 1 and 2 and Watts Bar Nuclear Plant (WBN) Unit 1 - Thirty-day Response to NRC Bulletin 2003-02, "Leakage from Reactor Pressure Vessel Lower Head Penetrations and Reactor Coolant Pressure Boundary Integrity," dated August 21, 2003,* ADAMS Accession No. ML032671210, September 22, 2003.

24. Jordan, T. J. *Response to NRC Bulletin 2003-02, "Leakage From Reactor Pressure Vessel Lower Head Penetrations and Reactor Coolant Pressure Boundary Integrity,"* (South Texas Project Units 1 and 2), ADAMS Accession No. ML033140306, November 4, 2003.

25. Byme, Stephen A. *Response to NRC Bulletin 2003-02, "Leakage from Reactor Pressure Vessel Lower Head Penetrations and Reactor Coolant Pressure Boundary Integrity,"* (Virgil C. Summer Nuclear Station), ADAMS Accession No. ML032660525, September 19, 2003.

26. Christian, David A. *Virginia Electric and Power Company (Dominion) Surry Power Station (SPS) Units 1 and 2 Thirty-Day Response to NRC Bulletin 2003-02 (SPS Unit 2) Ninety-Day Response to NRC Bulletin 2003-02 (SPS Unit 1), "Leakage from Reactor Pressure Vessel Lower Head Penetrations and Reactor Coolant Pressure Boundary Integrity,"* ADAMS Accession No. ML032690994, September 22, 2003.

27. Muench, Richard A. *Response to NRC Bulletin 2003-02, "Leakage From Reactor Pressure Vessel Lower Head Penetrations and Reactor Coolant Pressure Boundary Integrity,"* (Wolf Creek Generating Station), ADAMS Accession No. ML032690957, September 19, 2003.

Licensees' Post Inspection 60-day Responses to NRC Staff

1. Mitchell, Timothy, G. *60-Day Report for ANO-1 Reactor Pressure Vessel Head Inspection for Refueling Outage 1R18 Arkansas Nuclear One, Unit 1*, ADAMS Accession No. ML041960457, July 12, 2004.

2. Pearce, L. William. *NRC Bulletin 2003-02 Lower Head Inspection 60-Day Report for 1R16*, (Beaver Valley Power Station, Unit No. 1), ADAMS Accession No. ML050040183, December 29, 2004.

3. Pearce, L. William. *NRC Bulletin 2003-02 BV-2 Lower Head Inspection 60-Day Report*, (Beaver Valley Power Station, Units 1 and No. 2), ADAMS Accession No. ML033430288, December 2, 2003.

4. Joyce, T. P. *Braidwood Station, Unit 1 Sixty-Day Response to NRC Bulletin 2003-02, "Leakage from Reactor Pressure Vessel Lower Head Penetrations and Reactor Coolant Pressure Boundary Integrity,"* ADAMS Accession No. ML043520121, December 13, 2004.

5. Pacilio, Michael J. *Braidwood Station, Unit 2 Sixty-Day Response to NRC Bulletin 2003-02, "Leakage from Reactor Pressure Vessel Lower Head Penetrations and Reactor Coolant Pressure Boundary Integrity,"* ADAMS Accession No. ML040210555, January 15, 2004.

6. Kuczynski, Stephen E. *Byron Station Unit 1 Sixty-Day Response to NRC Bulletin 2003-02, "Leakage From Reactor Pressure Vessel Lower Head Penetrations and Reactor Coolant Pressure Boundary Integrity,"* ADAMS Accession No. ML033510558, December 12, 2003.

7. Kuczynski, Stephen E. *Byron Station Unit 2 Sixty-Day Response to NRC Bulletin 2003-02, "Leakage From Reactor Pressure Vessel Lower Head Penetrations and Reactor Coolant Pressure Boundary Integrity,"* ADAMS Accession No. ML041730450, June 2, 2004.

8. Young, Keith, D. *Response to Information Requested 60 Days After the Next Refueling Outage by NRC Bulletin 2003-02 and First Revised NRC Order EA-03-009*, (Callaway Plant Unit 1), ADAMS Accession No. ML042380494, August 12, 2004.

9. Jamil, D.M. *Response to NRC Bulletin 2003-02: Leakage from Reactor Pressure Vessel Lower Head Penetrations and Reactor Coolant Pressure Boundary Integrity*, (Catawba Nuclear Station Unit 1), ADAMS Accession No. ML040690711, February 26, 2004.

10. Jamil, D.M. *Response to NRC Bulletin 2003-02: Leakage from Reactor Pressure Vessel Lower Head Penetrations and Reactor Coolant Pressure Boundary Integrity*, (Catawba Nuclear Station Unit 2), ADAMS Accession No. ML043630309, December 16, 2004.

11. Zwolinski, John A. *Donald C. Cook Nuclear Plant Unit 1 Reactor Pressure Vessel Lower Head Penetration Inspection Results*, ADAMS Accession No. ML040920397, March 25, 2004.

12. Zwolinski, John A. *Donald C. Cook Nuclear Plant Unit 2, Unit 2 Reactor Pressure Vessel Upper and Lower Head Inspection Results*, ADAMS Accession No. ML050140187, January 6, 2005.

13. Madden, Fred W. *Comanche Peak Steam Electric Station, 60-Day Response Regarding NRC Bulletin 2003-02 and NRC Order EA.03-009*, (Comanche Peak, Unit 1), ADAMS Accession No. ML041910385, July 2, 2004.

14. Walker, Roger D. *Comanche Peak Steam Electric Station (CPSES) 60-Day Response Regarding NRC Bulletin 2003-02, "Leakage From Reactor Pressure Vessel Lower Head Penetrations and Reactor Coolant Pressure Boundary Integrity" and Report on RCS Conoseal Leakage*, (Comanche Peak, Unit 2), ADAMS Accession No. ML033640598, December 18, 2003.

15. Roderick, Daniel L. *Crystal River Unit 3 - 60 Day Report Regarding NRC Bulletin 2003-02, "Leakage from Reactor Pressure Vessel Lower Head Penetrations and Reactor Coolant Pressure Boundary Integrity,"* ADAMS Accession No. ML033570294, December 17, 2003.

16. Womack, Lawrence, F. *Diablo Canyon Unit 1, 30-Day Response to NRC Bulletin 2002-02, "Reactor Pressure Vessel Head and Vessel Head Penetration Nozzle Inspection Program, " 60-Day Response to NRC Bulletin 2003-02, "Leakage From Reactor Pressure Vessel Lower Head Penetrations and Reactor Coolant Pressure Boundary Integrity," and 60-Day Response to Revision 1 of NRC Order EA-03-009, "Issuance of First Revised NRC Order (EA-03-009) Establishing Interim Inspection Requirements for Reactor Pressure Vessel Heads at Pressurized Water Reactors,"* ADAMS Accession No. ML041950364, July 6, 2004.

17. Jacobs, Donna *60-Day Response to NRC Bulletin 2003-02, "Leakage From Reactor Pressure Vessel Lower Head Penetrations and Reactor Coolant Pressure Boundary Integrity," NRC Bulletin 2004-01, Inspection of Alloy 82/182/600 Materials Used in the Fabrication of Pressurizer Penetrations and Steam Space Piping Connections at Pressurized Water Reactors," and to Revision I of NRC Order EA-03-009. "Issuance of First Revised NRC Order (EA-03-009) Establishing Interim Inspection Requirements for Reactor Pressure Vessel Heads at Pressurized Water Reactors,"* Diablo Canyon, Unit 2, ADAMS Accession No. ML050530222, February 8, 2005.

18. Stinson, L. M. *Joseph M. Farley Nuclear Plant - Unit 2 Results of Reactor Pressure Vessel Head Inspections Required by First Revised NRC Order EA-03-009*, ADAMS Accession No. ML041490093, May 26, 2004.

19. Mecredy, Robert C. *60 Day Post Inspection Response to Bulletin 2003-02, Leakage from Reactor Pressure Vessel Lower Head Penetrations and Reactor Coolant Pressure Boundary Integrity R. E. Ginna Nuclear Power Plant*, ADAMS Accession No. ML033510022, December 9, 2003.

20. Morton, Terry C. *60-Day Summary Report, NRC Bulletin 2003-02, Leakage from Reactor Pressure Vessel Lower Head Penetrations and Reactor Coolant Pressure Boundary Integrity, Request (2),* (Shearon Harris Nuclear Power Plant, Unit No. 1), ADAMS Accession No. ML050260323, January 13, 2005.

21. Dacimo, Fred R. *Reactor Vessel Lower Head Inspection Results; Indian Point 2, Fall 2004 Refueling Outage (2R16)*, ADAMS Accession No. ML050260200, January 17, 2005.

22. Dacimo, Fred R. *Reactor Vessel Lower Head Inspection Results; Indian Point Unit 3, Spring 2005 Refueling Outage (3R13)*, ADAMS Accession No. ML051590184, May 31, 2005.

23. Coutu, Thomas *Nuclear Regulatory Commission Bulletin 2003-02: Leakage from Reactor Pressure Vessel Lower Head Penetrations and Reactor Coolant Pressure Boundary Integrity-Results of Inspections Conducted During Kewaunee Nuclear Power Plant Refueling Outage R-27*, ADAMS Accession No. ML050190262, January 7, 2005.

24. Peterson, G. R. *Response to NRC Bulletin 2003-02 Leakage from Reactor Pressure Vessel Lower Head Penetrations and Reactor Coolant Pressure Boundary Integrity,* (McGuire Nuclear Station Unit 1), ADAMS Accession No. ML042470040, June 9, 2004.

25. Peterson, G. R. *Response to NRC Bulletin 2003-02 Leakage from Reactor Pressure Vessel Lower Head Penetrations and Reactor Coolant Pressure Boundary Integrity,* (McGuire Nuclear Station Unit 2), ADAMS Accession No. ML033450391, December 4, 2003.

26. Hartz L. N. *Millstone Power Station Unit 3, Sixty-Day Response to NRC Bulletin 2003-02 Leakage from Reactor Pressure Vessel Lower Head Penetrations and Reactor Coolant Pressure Boundary Integrity*, ADAMS Accession No. ML041830261, June 24, 2004.

27. Hartz L. N. *North Anna Power Station Unit 1 Sixty-Day Response to NRC Bulletin 2003-02 Leakage from Reactor Pressure Vessel Lower Head Penetrations and Reactor Coolant Pressure Boundary Integrity*, ADAMS Accession No. ML043150052, November 9, 2004.

28. Hartz L. N. *North Anna Power Station Unit 2 Sixty-Day Response to NRC Bulletin 2003-02 Leakage from Reactor Pressure Vessel Lower Head Penetrations and Reactor Coolant Pressure Boundary Integrity*, ADAMS Accession No. ML041830264, June 24, 2004.

29. Jones, R. A. *Response to NRC Bulletin 2003-02: "Leakage from Reactor Pressure Vessel Lower Head Penetrations and Reactor Coolant Pressure Boundary Integrity,"* (Oconee Nuclear Station, Unit 1), ADAMS Accession No. ML040420217, February 3, 2004.

30. Jones, R. A. Response to NRC Bulletin 2003-02: "Leakage from Reactor Pressure Vessel Lower Head Penetrations and Reactor Coolant Pressure Boundary Integrity," (Oconee Nuclear Station, Unit 2), ADAMS Accession No. ML041890368, July 1, 2004.

31. Jones, R. A. *Response to NRC Bulletin 2003-02: "Leakage from Reactor Pressure Vessel Lower Head Penetrations and Reactor Coolant Pressure Boundary Integrity,"* (Oconee Nuclear Station, Unit 3), ADAMS Accession No. ML050270157, January 14, 2005.

32. Mauldin, David *APS' 60-Day After Plant Restart Letter in Response to NRC Bulletin 2003-02, Commitment No. 3 and First Revised NRC Order EA-03-009, Item IV.E-UIR11,* (Palo Verde Nuclear Generating Station Unit 1), ADAMS Accession No. ML042030316, July 9, 2004.

33. Mauldin, David *APS' 60-Day Letter in Response to NRC Bulletin 2003-02, Commitment No. 2,* (Palo Verde Nuclear Generating Station Unit 2), ADAMS Accession No. ML040280478, January 21, 2004.

34. Mauldin, David *APS' 60-Day Letter in Response to NRC Bulletin 2003-02, Commitment No. 2,* (Palo Verde Nuclear Generating Station Unit 3), ADAMS Accession No. ML050250214, January 11, 2005.

35. Koehl, Dennis L. *60-Day Report Pursuant to NRC Bulletin 2003-02 and NRC First Revised Order EA-03-009 for Unit 1 Refueling Outage 28 Reactor Vessel Inspections,* (Point Beach Nuclear Plant, Unit I), ADAMS Accession No. ML042300427, August 6, 2004.

36. Cayla, A. J. *60-Day Report Pursuant to NRC Bulletin 2003-02 and NRC Order EA-03-009 for Point Beach Nuclear Plant Unit 2 Reactor Vessel Inspections,* (Point Beach Nuclear Plant Unit 2), ADAMS Accession No. ML040230698, January 15, 2004.

37. Solymossy, Joseph M. *60-Day Report Pursuant to NRC Bulletin 2003-02 for 2004 Prairie Island Unit 1 Lower Head Penetration Inspection*, ADAMS Accession No. ML050330396, January 24, 2005.

38. Solymossy, Joseph M. *60-Day Report Pursuant to NRC Bulletin 2003-02 for 2003 Prairie Island Unit 2 Lower Head Penetration Inspection*, ADAMS Accession No. ML033500301, December 9, 2003.

39. Lucas, J. F. *Summary of Reactor Pressure Vessel Lower Head Inspection in Accordance with NRC Bulletin 2003-02, "Leakage from Reactor Pressure Vessel Lower Head Penetrations and Reactor Coolant Pressure Boundary Integrity,"* (H. B. Robinson, Unit No. 2), ADAMS Accession No. ML042110375, July 22, 2004.

40. Carlin, John *NRC Bulletin 2003-02 Inspection Results Leakage from Reactor Pressure Vessel Lower Head Penetrations and Reactor Coolant Pressure Boundary Integrity Salem Generating Station Unit 1*, ADAMS Accession No. ML041970268, July 6, 2004.

41. Carlin, John *Response to NRC Bulletin 2003-02 Leakage from Reactor Pressure Vessel Lower Head Penetrations and Reactor Coolant Pressure Boundary Integrity Salem Generating Station Unit 2*, ADAMS Accession No. ML040050267, December 23, 2003.

42. Warner, Mark E. *Seabrook Station Summary of Inspections of Reactor Pressure Vessel Lower Head Penetrations*, ADAMS Accession No. ML033630577, December 18, 2003.

43. Smith, James D. *Sequoyah Nuclear Plant (SQN) UNIT 2, NRC Bulletin 2003-02, "Leakage from Reactor Pressure Vessel Lower Head Penetrations and Reactor Coolant Pressure Boundary Integrity," and NRC Order EA-03-009 - Interim Inspection Requirements for Reactor Pressure Vessel Heads at Pressurized Water Reactors*, ADAMS Accession No. ML040580079, February 5, 2004.

44. Jordan, T. J. *South Texas Project Unit 1 Docket No. STN 50-498 60-Day Response to NRC Bulletin 2003-02, "Leakage from Reactor Pressure Vessel Lower Head Penetrations and Reactor Coolant Pressure Boundary Integrity,"* ADAMS Accession No. ML051610219, June 2, 2005.

45. Jordan, T. J. *South Texas Project Unit 2, 60-Day Response to NRC Bulletin 2003-02, "Leakage from Reactor Pressure Vessel Lower Head Penetrations and Reactor Coolant Pressure Boundary Integrity,"* ADAMS Accession No. ML041770334, June 17, 2004.

46. Byrne, Stephen A. *60 Day Follow-Up Letter to NRC Bulletin 2003-02, Leakage from Reactor Pressure Vessel Lower Head Penetrations and Reactor Coolant Pressure Boundary Integrity,* (Virgil C. Summer Nuclear Station), ADAMS Accession No. ML040270204, January 23, 2004.

47. Hartz, L. N. *Sixty-Day Response to NRC Bulletin 2003-02, Leakage from Reactor Pressure Vessel Lower Head Penetrations and Reactor Coolant Pressure Boundary Integrity,* (Surry Power Station Unit 1), ADAMS Accession No. ML050190177, January 18, 2005.

48. Hartz, L. N. *Sixty-Day Response to NRC Bulletin 2003-02, Leakage from Reactor Pressure Vessel Lower Head Penetrations and Reactor Coolant Pressure Boundary Integrity,* (Surry Power Station Unit 2), ADAMS Accession No. ML040420238, January 30, 2004.

49. Gallagher, Michael P. *TMI Unit 1 Sixty-Day Response to NRC Bulletin 2003-02, "Leakage from Reactor Pressure Vessel Lower Head Penetrations and Reactor Coolant*

Pressure Boundary Integrity," (Three Mile Island, Unit 1), ADAMS Accession No. ML040300250, January 22, 2004.

50. Jones, Terry O. *NRC Bulletin 2003-02 Reactor Pressure Vessel Lower Head Penetrations Post Outage Inspection Results*, (Turkey Point Unit 3), ADAMS Accession No. ML050350140, January 27, 2005.

51. Jones, Terry O. *NRC Bulletin 2003-02 Leakage from Reactor Pressure Vessel Lower Head Penetrations and Reactor Coolant Pressure Boundary Integrity Inspection Results*, (Turkey Point Unit 4), ADAMS Accession No. ML033640596, December 19, 2003.

52. Gasser, Jeffrey T. *Vogtle Electric Generating Plant Results of Reactor Pressure Vessel Head Inspections Required by NRC Bulletin 2003-02*, (Vogtle Electric Generating Plant, Unit 1), ADAMS Accession No. ML033570131, December 19, 2003.

53. Gasser, Jeffrey T. *Vogtle Electric Generating Plant Results of Reactor Pressure Vessel Head Inspections Required by First Revised NRC Order EA-03-009 and NRC Bulletin 2003-02*, (Vogtle Electric Generating Plant, Unit 2), ADAMS Accession No. ML041830360, June 28, 2004.

54. Pace, P. L. *Watts Bar Nuclear Plant (WBN) Unit 1 - NRC Order EA-03-009 - Interim Inspection Requirements for Reactor Pressure Vessel Heads and Bulletin 2003-02 - Leakage from Reactor Pressure Vessel Lower Head Penetrations and Reactor Coolant Pressure Boundary Integrity,* ADAMS Accession No. ML033500135, December 10, 2003.

55. Muench, Richard A. 60-Day Report for NRC Bulletin 2003-02: 'Leakage from Reactor Pressure Vessel Lower Head Penetrations and Reactor Coolant Pressure Boundary Integrity," (Wolf Creek), ADAMS Accession No. ML040300234, January 22, 2004.

Supplemental Correspondence

1. Hartz, L. N. *MRP 2003-017 Recommendations for PWR Owners with Alloy 600 Bottom Mounted Reactor Vessel Instrument Nozzles,* ADAMS Accession No. ML031920395, June 23, 2003.

2. Beasley, J. B. *Joseph M. Farley Nuclear Plant - Unit 1, Results of Reactor Pressure Vessel Head Inspections Required by Order EA-03-009,* ADAMS Accession No. ML031830334, June 30, 2003.

3. Sheppard, J. J. *South Texas Project, Units 1 and 2, Docket Nos. STN 50-498, STN 50-499, Supplement to STP Commitment to Investigate and Repair Bottom Mounted Instrumentation Penetration Indications,* ADAMS Accession No. ML032020116, July 17, 2003.

4. Bezilla, Mark B. *Davis-Besse Nuclear Power Station Incore Monitoring Instrumentation Nozzle Inspections,* ADAMS Accession No. ML032160384, July 30, 2003.

5. Birmingham, Joseph L. *Summary of November 25, 2003, Meeting with Industry and the Materials Reliability Project on Long-Term Inspection Plans for Bottom Mounted Nozzles,* ADAMS Accession No. ML033520439, December 18, 2003.

6. Mauldin, David *Palo Verde Nuclear Generating Station (PVNGS), Unit 2, Docket No. STN 50-529, Unit 2's 60-Day After Plant Restart Letter in Response to NRC Bulletin 2003-02, Commitment No. 3 and Order EA-03-009, Item IV.E.,* ADAMS Accession No. ML040550543, February 17, 2004.

7. Hartz, L. N. *MRP 2004-04 PWR Owners with Alloy 600 Bottom Mounted Instrument Nozzles,* ADAMS Accession No. ML041480017, May 14, 2004.

8. Jordan, T. J. *South Texas Project, Units 1 and 2, Docket Nos. STN 50-498, STN 50-499, Corrected Response to NRC Bulletin 2003-02,* ADAMS Accession No. ML041620146, June 7, 2004.

9. Monarque, Stephen R. Summary of April 22, 2004, Conference Call Regarding Duke Power Company's Response to NRC Bulletin 2003-02, "Leakage from Reactor Pressure Vessel Lower Head Penetrations and Reactor Coolant Pressure Boundary Integrity," for Catawba Nuclear Station, Unit 1, ADAMS Accession No. ML041910016, July 1, 2004.

10. Stall, J. A. *Revised Response to NRC Bulletin 2003-02 Leakage from Reactor Pressure Vessel Lower Head Penetrations and Reactor Coolant Pressure Boundary Integrity,* (Turkey Point, Units 3 and 4), ADAMS Accession No. ML042110162, July 27, 2004.

11. Birmingham, Joseph L. *Summary of July 19, 2004, Teleconference with Industry and the Materials Reliability Project on Safety Assessment for Bottom Mounted Nozzles,* ADAMS Accession No. ML042160490, August 3, 2004.

12. Donohew, Jack *Callaway Plant, Unit 1 - Summary of Conference Call on July 28, 2004, to Discuss Volumetric Examinations of Reactor Vessel Bottom Mounted Nozzles (TAC No. MC0527),* ADAMS Accession No. ML042450744, September 1, 2004.

13. Bezilla, Mark B. *Davis-Besse Nuclear Power Station (DBNPS), Mid-Cycle 14 Outage Reactor Head Visual Inspection Results (Confirmatory Order EA-03-214)*, ADAMS Accession No. ML050350194, February 3, 2005.

14. Korsnick, Mary G. *R.E. Ginna Nuclear Power Plant, Docket No. 50-244, Response to First Revised Order EA-03-009 and Bulletin 2003-02*, ADAMS Accession No. ML051660326, June 10, 2005.

15. Mauldin, David *Palo Verde Nuclear Generating Station (PVNGS), Unit 2, Docket No. STN 50-529, APS' 60-Day After Plant Restart Letter in Response to First Revised NRC Order EA-03-009, Item IV.E, NRC Bulletin 2003-02, Commitment No. 3 and NRC Bulletin 2004-01, Commitment No. 2 - U2R12*, ADAMS Accession No. ML052070715, July 18, 2005.

16. Birmingham, Joseph L. *Summary of September 29, 2005, Meeting with Industry and the Materials Reliability Project on Long-Term Inspection Plans for Bottom Mounted Nozzles*, ADAMS Accession No. ML052780038, October 18, 2005.

17. Correspondence from EPRI to the NRC dated April, 17, 2006–Information Regarding Materials and Fabrication Methods used for the Bottom Mounted Instrumentation Penetrations, ADAMS Accession No. ML061100063, April 17, 2006.

18. Nuclear Energy Institute, NEI-03-08, "Guideline for the Management of Materials Issues," May 2003, ADAMS Accession No. ML032190048.

19. Correspondence from Project Manager, EPRI MRP Assessment ITG to the MRP Technical Advisory Group dated September 15, 2006, "Bottom Mounted Nozzle Inspection Summary (Summer 2006 Update)," ADAMS Accession No. ML062620235, September 19, 2006.

NRC Inspection Reports

1. NRC Inspection Manual, Temporary Instruction 2515/152, Revision 1, *Reactor Pressure Vessel Lower Head Penetration Nozzles (NRC Bulletin 2003-02)* ADAMS Accession No. ML033230229, November 5, 2003.

2. Arkansas Nuclear One - NRC Integrated Inspection Report 05000313/2004003 and 05000368/2004003, ADAMS Accession No. ML042100540, July 27, 2004.

3. Beaver Valley Power Station - NRC Integrated Inspection Report 05000334/2004006 and 05000412/2004006, ADAMS Accession No. ML050320284, January 31, 2005.

4. Beaver Valley Power Station - NRC Integrated Inspection Report 05000334/2003004 and 05000412/2003004, ADAMS Accession No. ML033140587, November 10, 2003.

5. Braidwood Station, Units 1 and 2 - NRC Integrated Inspection Report 05000456/2004008 and 05000456/2004008, ADAMS Accession No. ML050270180, January 26, 2005.

6. Braidwood Station, Units 1 and 2 - NRC Integrated Inspection Report 05000456/2004008 and 05000457/2004008, ADAMS Accession No. ML040280151, January 26, 2005.

7. Byron Station, Units 1 and 2 - NRC Integrated Inspection Report 05000454/2003007 and 05000455/2003007, ADAMS Accession No. ML040280213, January 26, 2004.

8. Byron Station, Units 1 and 2 - NRC Integrated Inspection Report 05000454/2004004 and 05000455/2004004, ADAMS Accession No. ML042050022, July 20, 2004.

9. Byron Station, Units 1 and 2 - NRC Integrated Inspection Report 05000454/2003007 and 05000455/2003007, ADAMS Accession No. ML040280213, January 26, 2004.

10. Callaway Plant - NRC Integrated Inspection Report 05000483/2004003, ADAMS Accession No. ML041290222, August 5, 2004.

11. Catawba Nuclear Station- NRC Integrated Inspection Report 05000413/2003005 and 05000414/2003005, ADAMS Accession No. ML040200725, January 16, 2004.

12. Catawba Nuclear Station- NRC Integrated Inspection Report 05000413/2004005 and 05000414/2004005, ADAMS Accession No. ML050680037, October 15, 2004.

13. D. C. Cook Nuclear Power Plant, Units 1 and 2 NRC Integrated Inspection Report 05000315/2003012 and 05000316/2003012, ADAMS Accession No. ML040210391, January 20, 2004.

14. D. C. Cook Nuclear Power Plant, Units 1 and 2 NRC Integrated Inspection Report 05000315/2004012 and 05000316/2004012, ADAMS Accession No. ML050200039, January 19, 2005.

15. Comanche Peak Steam Electric Station - NRC Integrated Inspection Report 05000445/2004003 and 05000446/2004003, ADAMS Accession No. ML042090544, July 26, 2004.

16. Comanche Peak Steam Electric Station - NRC Integrated Inspection Report 05000445/2003004 and 05000446/2003004, ADAMS Accession No. ML040220055, January 20, 2004.

17. Crystal River, Unit 3 - NRC Integrated Inspection Report 05000302/2003006, and Exercise of Enforcement Discretion, ADAMS Accession No. ML040270008, January 26, 2004.

18. Diablo Canyon Power Plant - NRC Integrated Inspection Report 05000275/2004003, and 05000323/2004003, ADAMS Accession No. ML042250352, August 12, 2004.

19. Diablo Canyon Power Plant - NRC Integrated Inspection Report 05000275/2004005, and 05000323/2004005, ADAMS Accession No. ML050450591, February 11, 2005.

20. Joseph M. Farley Nuclear Plant - NRC Integrated Inspection Report 05000348/2004005, and 05000364/2004005, ADAMS Accession No. ML050280343, January 27, 2005.

21. Joseph M. Farley Nuclear Plant - NRC Integrated Inspection Report 05000348/2004002, and 05000364/2004002, ADAMS Accession No. ML041130520, April 22, 2004.

22. R. E. Ginna Nuclear Power Plant - NRC Integrated Inspection Report 05000244/2003006, ADAMS Accession No. ML033140029, November 6, 2003.

23. Shearon Harris Nuclear Power Plant - NRC Integrated Inspection Report 05000400/2004006, ADAMS Accession No. ML050310409, January 28, 2005.

24. Indian Point Nuclear Generating Unit 2 - NRC Integrated Inspection Report No. 05000247/2004012, ADAMS Accession No. ML050340248, February 2, 2005.

25. Indian Point Nuclear Generating Unit 3 - NRC Integrated Inspection Report No. 05000286/2005002, ADAMS Accession No. ML051320163, May 12, 2005.

26. Kewaunee Nuclear Power Plant - NRC Integrated Inspection Report 05000305/2004009, ADAMS Accession No. ML050460220, February 14, 2005.

27. McGuire Nuclear Station - NRC Integrated Inspection Report 05000369/2003005, and 05000370/2004003, ADAMS Accession No. ML041040199, April 12, 2004.

28. McGuire Nuclear Station - NRC Integrated Inspection Report 05000369/2003005, and 05000370/2003005, ADAMS Accession No. ML040130636, January 12, 2004.

29. Millstone Power Station, Unit 2 and Unit 3 - NRC Integrated Inspection Report 05000336/2004006, and 05000423/2004006, ADAMS Accession No. ML042110368, July 29, 2004.

30. North Anna Power Station - NRC Integrated Inspection Report 05000338/2004003, and 05000339/20040003, ADAMS Accession No. ML042050510, July 23, 2004.

31. North Anna Power Station - NRC Integrated Inspection Report 05000338/2004005, 05000339/2004005, and 07200016/2004002, ADAMS Accession No. ML042960497, October 21, 2004.

32. Oconee Nuclear Station - Integrated Inspection Report 05000269/2003005, 05000270/2003005, and 05000287/2003005 and Independent Spent Fuel Storage Installation Inspection Report 72-04/2003001, ADAMS Accession No. ML040270297, January 26, 2004.

33. Oconee Nuclear Station - Final Significance Determination - Integrated Inspection Report 05000269/2004003, 05000270/2004003, and 05000287/2004003, ADAMS Accession No. ML042050511, July 23, 2004.

34. Oconee Nuclear Station -- Integrated Inspection Report 05000269/2004005, 05000270/20040005, and 05000287/2004005, ADAMS Accession No. ML050280392, January 27, 2005.

35. Palo Verde Nuclear Generating Station - NRC Integrated Inspection Report 05000528/2004003, 05000529/2004003, and 05000530/2004003, ADAMS Accession No. ML042220267, August 9, 2004.

36. Palo Verde Nuclear Generating Station - NRC Integrated Inspection Report 05000528/2003005, 05000529/2003005, and 05000530/2003005, ADAMS Accession No. ML040300961, January 30, 2004.

37. Palo Verde Nuclear Generating Station - NRC Integrated Inspection Report 05000528/2004005, 05000529/2004005, and 05000530/2004005, ADAMS Accession No. ML050390475, February 8, 2005.

38. Point Beach Nuclear Plant, Units 1 and 2 - NRC Integrated Inspection Report 05000266/2004003; 05000301/2004003, ADAMS Accession No. ML042260093, August 12, 2004.

39. Point Beach Nuclear Plant, Units 1 and 2 - NRC Integrated Inspection Report 05000266/2003009; 05000301/2003009, ADAMS Accession No. ML040340170, January 30, 2004.

40. Prairie Island Nuclear Generating Plant, Units 1 and 2 - NRC Integrated Inspection Report 05000282/2004008; 05000306/2004008, ADAMS Accession No. ML050280336, January 27, 2005.

41. Prairie Island Nuclear Generating Plant, Units 1 and 2 - NRC Integrated Inspection Report 05000282/2003005; 05000306/2003005, ADAMS Accession No. ML033040227, October 29, 2003.

42. H. B. Robinson Nuclear Power Plant - NRC Integrated Inspection Report 05000261/2004003, ADAMS Accession No. ML041940342, July 9, 2004.

43. Salem Nuclear Generating Station - NRC Integrated Inspection Report 05000272/2004003 and 05000311/2004003, ADAMS Accession No. ML042250128, August 11, 2004.

44. Salem Nuclear Generating Station - NRC Integrated Inspection Report 05000272/2003009 and 05000311/2003009, ADAMS Accession No. ML040440024, February 12, 2004.

45. Seabrook Station - NRC Integrated Inspection Report 05000443/2003006, ADAMS Accession No. ML040230168, January 23, 2004.

46. Sequoyah Nuclear Power Plant - NRC Integrated Inspection Report 05000327/2004005, and 05000328/2004005, ADAMS Accession No. ML050250002, January 24, 2005.

47. Sequoyah Nuclear Power Plant - NRC Integrated Inspection Report 05000327/2003006, and 05000328/2003006, ADAMS Accession No. ML040270032, January 26, 2004.

48. Virgil C. Summer Nuclear Station - NRC Integrated Inspection Report No. 05000395/2003005, ADAMS Accession No. ML040270300, January 26, 2004.

49. South Texas Project Electric Generating Station - NRC Integrated Inspection Report 05000498/2005002 and 05000499/2005002, ADAMS Accession No. ML051290315, May 9, 2005.

50. South Texas Project Electric Generating Station - NRC Integrated Inspection Report 05000498/2004003 and 05000499/2004003, ADAMS Accession No. ML050390020, August 5, 2004.

51. Surry Power Station - NRC Integrated Inspection Report Nos. 05000280/2004005 and 05000281/2004005, ADAMS Accession No. ML050280264, January 27, 2005.

52. Surry Power Station - NRC Integrated Inspection Report Nos. 05000280/2003005 and 05000281/2003005, ADAMS Accession No. ML040280056, January 26, 2004.

53. Three Mile Island Station, Unit 1 - NRC Integrated Inspection Report 05000289/2003005, ADAMS Accession No. ML040230730, January 23, 2004.

54. Turkey Point Nuclear Plant - Integrated Inspection Report 05000250/2004005 and 05000251/2004005, ADAMS Accession No. ML050280368, January 28, 2005.

55. Turkey Point Nuclear Plant - Integrated Inspection Report 05000250/2003005 and 05000251/2003005, ADAMS Accession No. ML040220017, January 21, 2004.

56. Vogtle Electric Generating Plant - NRC Integrated Inspection Report 05000424/2003005 and 05000425/2003005 (NOED NO. 03-6-004), ADAMS Accession No. ML040260029, January 23, 2004.

57. Vogtle Electric Generating Plant - NRC Integrated Inspection Report 05000424/2004004 and 05000425/2004004, ADAMS Accession No. ML042050159, July 23, 2004.

58. Watts Bar NRC Integrated Inspection Report 05000390/2003004 and 05000391/2003004, ADAMS Accession No. ML033010084, October 27, 2003.

59. Wolf Creek Generating Station - NRC Integrated Inspection Report 05000482/2003006, ADAMS Accession No. ML040340162, February 3, 2004.